I0051151

Iron and
Manganese Removal
Handbook

Second Edition

**American Water Works
Association**

Iron and Manganese Removal Handbook, Second Edition
Copyright © 1999, 2015 American Water Works Association

All rights reserved. No part of this publication may be reproduced or transmitted in any form or by any means, electronic or mechanical, including photocopy, recording, or any information or retrieval system, except in the form of brief excerpts or quotations for review purposes, without the written permission of the publisher.

Disclaimer
This book is provided for informational purposes only, with the understanding that the publishers, editors, and authors are not thereby engaged in rendering engineering or other professional services. The publishers, editors, and authors make no claim as to the accuracy of the book's contents, or their applicability to any particular circumstance, and accept no liability to any person for the information provided in this program, or for the loss or damages incurred by any person as a result of reliance on its contents. The reader is urged to consult with an appropriate licensed professional before taking any action or making any interpretation that is within the realm of a licensed professional practice.

AWWA Senior Manager of Editorial Development and Production: Gay Porter De Nileon
AWWA Senior Production Editor: Cheryl Armstrong

Library of Congress Cataloging-in-Publication Data
Civardi, John, author.
  Iron and manganese removal handbook / by John Civardi and Mark Tompeck. -- Second edition.
      pages cm
  Includes bibliographical references.
  ISBN 978-1-58321-985-0
  1. Water--Purification--Iron removal--Handbooks, manuals, etc. 2. Water--Purification--Manganese removal--Handbooks, manuals, etc. I. Tompeck, Mark, author. II. Title.
  TD466.C58 2015
  628.1'68--dc23
                            2015033293

Printed in the United States of America

ISBN: 978-1-58321-985-0
eISBN: 978-1-61300-263-6

American Water Works Association

6666 W. Quincy Avenue
Denver, CO 80235-3098
800.926.7337
www.awwa.org

Contents

Figures

Tables

Acknowledgments

We are grateful to the many people who contributed in a variety of ways to the creation of this book. First and foremost, we thank Elmer O. Sommerfeld, the author of the previous edition of the *Iron and Manganese Removal Handbook*, for laying the foundation for a practical tool in addressing the problem of iron and manganese in drinking water.

Many of our colleagues at Hatch Mott MacDonald reviewed chapters and made helpful suggestions, including Margie Gray, John Schneekloth, and James Poirier. Don Nusser, Sustainability Practice Leader, prepared much of the information in chapter 2 and contributed to our thinking about the sustainability issues posed by iron and manganese removal. Geoff Wisner provided valuable editorial assistance at every step of the way. William Bertera of the Institute for Sustainable Infrastructure provided assistance on sustainability issues. We are extremely grateful for the assistance of Dr. Taha Marhaba, Professor and Chairman of the John A. Reif, Jr., Department of Civil and Environmental Engineering of the New Jersey Institute of Technology. Dr. Marhaba provided valuable assistance regarding the fundamental chemical and hydraulic aspects associated with iron and manganese removal.

We are grateful as well to the numerous suppliers who contributed to this book, including Jackie Cutter of Anthratech Western, Raymond Jones of Hungerford & Terry, Chris Savino of Layne, Chris Hanson of Meurer Research, Chris Wilkinson of NO-DES, Ana Van Den Hende of PSI Process and Equipment, and Tom Perry of Veolia Water.

Any errors in this book remain the responsibility of the authors. We hope you find this new edition of the *Iron and Manganese Removal Handbook* useful and interesting.

Foreword

During more than 30 years as a water professional in the Pacific Northwest, I found that few things were more alarming to customers than seeing red water or black water come out of their faucets. Two such contaminants that can cause colored water are iron and manganese. Although iron and manganese in drinking water are generally not hazardous to health, they are unacceptable to consumers, and removing them is a multimillion-dollar challenge.

For years, operators and engineers have relied on the *Iron and Manganese Removal Handbook*. First published in 1984 by the Saskatchewan Environment and Resource Management Department, the handbook was updated by Elmer O. Sommerfeld and published by the American Water Works Association in 1999.

Sixteen years later, there is a pressing need for an updated version of this valuable resource. AWWA has done the industry a great service in making this new edition of the *Iron and Manganese Removal Handbook* available.

The new edition emphasizes the science of iron and manganese contamination by offering firsthand field experience, detailed case studies, sustainability considerations, and practical evaluations of current and emerging methods for iron and manganese removal. It has been thoroughly updated by two leading water professionals: John Civardi and Mark Tompeck of Hatch Mott MacDonald.

During more than 25 years of experience in water treatment, John Civardi has been the lead project engineer for many groundwater treatment projects to remove iron and manganese. He has published papers in *Water World,* the *New England Water Works Association Journal*, the International Water Association conference proceedings, and corporate in-house publications. In 2003 he was the winner of the New England Water Works Association's Dexter Brackett Award for Best Paper.

As a senior vice president in Hatch Mott MacDonald's Water Division and the firm's water practice leader, Mark Tompeck has developed expertise in many areas of water supply engineering, from planning and design to permitting and construction engineering. He has served as project manager on a wide variety of projects involving the design and construction of treatment plants, pumping stations, chemical storage/feed facilities, water and sewer pipelines, storage tanks, and wells.

I am confident that the *Iron and Manganese Removal Handbook* will continue to be a go-to resource for engineers and treatment plant operators in North America and around the world.

Gregory DiLoreto, PE, F.ASCE
President 2013
American Society of Civil Engineers

Preface

Iron and manganese, elements 26 and 25 on the periodic table, find their way into drinking water when water percolates through soil and rock, dissolving minerals such as amphiboles, magnetite, and carbonates and holding them in solution.

Deep wells, where water is in prolonged contact with minerals, may yield high levels of iron and manganese. The iron and manganese levels in groundwaters tend to be relatively stable over time. Reservoirs and streams fed by reservoirs can experience elevated levels of iron and manganese, which can vary seasonally. The distribution system may be a source of iron due to corrosion or the growth of iron bacteria.

In the United States, the US Environmental Protection Agency (USEPA) has classified iron and manganese as a Secondary Standard that is nonhazardous to human health, although some studies have linked manganese to lower IQs in children and symptoms similar to Parkinson's disease in adults. However, iron and manganese are troublesome in other ways. Iron and manganese are known to adversely affect the aesthetics of the water at the customer's tap, and the aesthetic effects can lead to decreased consumer confidence and result in an increase in customer complaints.

Iron stains laundry, dishes, utensils, fixtures, and glassware reddish brown, while manganese stains them brownish black. Soap and detergent will not clean these stains, and chlorine bleach and alkaline builders may make them worse. Iron and manganese can react with the tannins in coffee, tea, and some alcoholic drinks to produce a black sludge. Iron can make water taste metallic and unpleasant for people and farm animals, while even smaller concentrations of manganese can be offensive. Vegetables cooked in iron-rich water can turn dark and be unappetizing.

Bacteria that feed on these metals form a slime inside toilet tanks that becomes reddish brown with iron or blackish brown with manganese. Bacterial slime may clog water systems and stain laundry and fixtures. Sediments and precipitates of iron and manganese build up and shorten the lives of pipelines, pressure tanks, hot-water heaters, and water softeners, and require more energy to pump water through narrowed pipes or to heat it with electric heating rods covered by deposits. The cost in energy, dollars, and consumer dissatisfaction can be considerable. In 2013 in Winnipeg, Manitoba, where high levels of iron and manganese have been recorded, the city received 1,344 complaints about discolored water in just three weeks (CBC News 2013).

The *Iron and Manganese Removal Handbook* is one of the leading resources for coping with these challenges. First published in 1984 by the Saskatchewan Environment and Resource Management Department, the handbook was revised by Elmer O. Sommerfeld and published by AWWA in 1999. We are pleased to have the opportunity to update this valuable resource using field experience, detailed case studies, practical evaluations of current and emerging methods for iron and manganese removal, and guidance on the handling of residuals.

John Civardi, PE
Hatch Mott MacDonald

Mark Tompeck, PE, DBIA
Practice Leader (Water), Hatch Mott MacDonald
New Jersey Director, American Water Works Association

Introduction

This book is an updated version of the *Iron and Manganese Removal Handbook*, published by the American Water Works Association in 1999.

The original handbook was developed in part as a practical training tool for operators and engineers. Over the past two decades, a number of issues have impacted the design and operation of iron and manganese treatment systems:

- Lower maximum contaminant levels (MCLs) for disinfection by-products
- Broader understanding of the impact of radionuclides on the performance of treatment systems
- Limited availability of glauconite greensand
- Desire to have systems that are sustainable
- Advancements in membrane technology
- Need to treat additional synthetic organic compounds and the impact of multiple treatment processes on the iron/manganese removal system
- Increased attention to distribution system water quality
- Increased use of oxidants other than traditional chlorine gas and 15 percent sodium hypochlorite

The purpose of this handbook is the same as the original handbook, which was developed as a training tool and as a useful reference for anyone interested in removing iron and manganese from potable water, as well as the basic operation of water treatment and filtration plants.

In this new edition, more detail is devoted to the selection and design of the treatment process and to minimizing the use of vendor-engineered packages that have become so common throughout the industry. Unit

1

processes are described and calculations are provided so that the reader can understand the impacts of changes to the design components.

This edition also attempts to address the question *why*. When approaching any water quality issue, the investigator may wish to consider why decisions are made and standards followed. For example, why is the secondary maximum contaminant level (SMCL) at 0.3 mg/L for iron? Why is chlorine the most common oxidant? Why is the traditional filter loading rate 3 gpm/ft^2? In an effort to increase the understanding of water professionals, this book is written with the *why* question in mind.

The handbook is organized into the following chapters:

Chapter 1. Assessing the Problem

Chapter 1 describes the occurrence of iron and manganese in drinking water systems. It presents an overview of applicable regulations and of the potential simultaneous compliance issues that may be impacted by iron and manganese treatment.

Chapter 2. Sustainability and Engineering Design

This chapter defines sustainability, examines the importance of the sustainable approach, considers how sustainability can be measured, and outlines core sustainability factors as they apply to the removal of iron and manganese.

Chapter 3. Basic Chemistry

This chapter explains the basic chemical reactions associated with iron and manganese. These chemical reactions affect the forms in which iron and manganese appear and the types of treatment that can be used to control and remove them.

Chapter 4. Overview of Treatment Technologies

This chapter provides the framework for the remaining chapters of the book and introduces the reader to the potential treatment options that are available to address iron and manganese issues. The chapter also describes a methodology for evaluating and selecting the optimum treatment system.

Chapter 5. Oxidation

In natural waters, iron and manganese rarely exist in particulate form. Therefore, oxidation is the first step in any treatment system. The selection of the optimum oxidation method(s) is traditionally based on efficiency, chemical and equipment cost, and the chemical characteristics of the source water. This chapter describes when to use which oxidation technologies and offers an overview of the design of the technology.

Chapter 6. Clarification

For source water with high iron concentrations (above 5–10 mg/L), filtration technologies generally result in a significant amount of backwash water and thus operate at low water efficiency due to short filter run times. In such cases, it may be desirable to provide clarification to remove the bulk of the iron/manganese to improve the filtration process. Clarification has traditionally been limited to sedimentation, and in some instances solids-contact clarification. Over the last two decades, advancements have been made to plate and tube settlers, which has made these high-rate systems viable treatment technologies. In some instances, high-rate technologies such as ballasted flocculation have also been considered. Chapter 6 presents practical design information regarding clarification technologies.

Chapter 7. Filtration

Filtration has undergone significant advancements in recent years. Historically, filtration was limited to either conventional sand/anthracite or glauconite. Over the past 20 years, however, glauconite supply has become limited and several alternative filtration products have been developed. Chapter 7 describes the design and operating features of the various types of alternate filter media and includes a discussion of biological filtration.

Chapter 8. Residuals

The iron/manganese treatment process generates residuals that require disposal. Treatment can be provided to reduce the volume of waste and improve the water efficiency of the entire treatment system. Chapter 8 describes residuals treatment options and practical design guidelines.

Chapter 9. Distribution System Water Quality

A significant number of customer complaints are prompted by dirty water, resulting from corrosion in metallic distribution systems and changes in treatment plant practices, such as lowering the chlorine residual on top of the filters to reduce the amount of disinfection by-products. This chapter presents an overview of distribution system water quality as it relates to iron and manganese and provides potential sampling/data collection and remediation options should a system exhibit elevated iron and manganese.

Chapter 10. Case Studies

These case studies present solutions to a variety of challenges in the field of iron and manganese removal. They include several design examples intended to demonstrate different treatment system designs.

Appendices

Appendix A details the results of pilot studies conducted on iron and manganese removal processes. Appendix B provides conversion tables and calculation tools for dosing and analyses.

1

Assessing the Problem

This chapter describes the occurrence of iron and manganese in drinking water systems. It presents an overview of applicable regulations and of the potential simultaneous compliance issues that may be impacted by iron and manganese treatment.

SOURCES OF IRON AND MANGANESE

Iron

Iron (Fe) is the second most abundant metal on earth, after aluminum. It accounts for about 5 percent of the earth's crust. Pure iron is rarely found in nature because the iron ions Fe^{2+} and Fe^{3+} readily combine with oxygen- and sulfur-containing compounds to form oxides, hydroxides, carbonates, and sulfides. Iron is most commonly found in nature in the form of its oxides (World Health Organization 2003).

The median iron concentration in rivers has been reported to be 0.7 mg/liter (mg/L). In anaerobic groundwater where iron is in the form of iron(II), concentrations will usually be 0.5 mg/L to 10 mg/L, but concentrations up to 50 mg/L can sometimes be found. Concentrations of iron in drinking water are normally less than 0.3 mg/L. Iron concentrations may be higher in countries where various iron salts are used as coagulating agents in water treatment plants, or where cast-iron, steel, and galvanized iron pipes are used for water distribution (WHO 2003).

Manganese

Manganese (Mn) is found throughout the environment and is needed for normal physiological functions in humans and animals. Exposure to low

5

levels of manganese in the diet is considered essential for human health. The average daily intake from food is 1–5 mg/day (USEPA 2010).

Manganese is detected in approximately 70 percent of groundwater sites in the United States (WHO 2011), but generally below levels of public health concern (USEPA 2002). Average manganese levels in drinking water are approximately 0.004 mg/L (USEPA 2010).

One survey of public water supplies found about 95 percent contained manganese at a concentration of less than 0.1 mg/L (Cooke 2014). The National Inorganic and Radionuclide Survey collected data from 989 community public water systems served by groundwater in 49 US states between 1984 and 1986 and found that manganese was detected in 68 percent of the groundwater systems, with a median concentration of 0.01 mg/L. Supplemental survey data from public water systems supplied by surface water in five states reported occurrence ranges similar to those of groundwater.

GROUNDWATER SOURCES

The iron and manganese found in groundwater are generally in a dissolved state and generally remain stable over time. While most groundwater does not contain microorganisms, in some aquifers the presence of iron and manganese promotes the growth of iron-reducing bacteria called crenoforms. These have scientific names, including *Crenothrix, Gallionella, Leptothrix,* and *Sphaerotilus.*

Crenoforms congregate in piping to form heavy, jelly-like, stringy masses that can impair the water-carrying capacity of an entire system. Allowing the formation of these organisms through inadequate removal of iron and manganese and inadequate disinfection of filtered water at the water treatment plant is almost sure to result in a substantial cost in time and money to flush and/or swab distribution lines. Crenoforms are also likely to reform following removal.

According to Cullimore (1993), "It was found during that research phase that 95 percent of the groundwaters tested in Saskatchewan, Canada, were positive for IRB (iron-reducing bacteria). Microscopic examination found that, of the sheathed and stalked IRB, *Crenothrix, Leptothrix, Sphaerotilus,* and *Gallionella* were frequently dominant types. Given the universal nature of the presence of IRB it becomes more critical to appreciate their relative aggressivity."

Cullimore continues, "The presence of IRB can result in elevated distribution system iron levels even if the raw water iron levels are less than 0.3 mg/L. For systems having problems with red water, consideration should be given to analyzing the water for IRB. The BART (Biological Activity Reaction Test) can be used to identify if a groundwater or distribution system has IRB." He notes that contaminants can be transported between groundwater and surface water when stream levels are low and aquifers serve as a source of recharge for a surface water.

SURFACE WATER SOURCES

Surface waters include reservoirs, rivers, creeks, and streams, including streams that are fed from upstream reservoirs. In flowing rivers and streams, iron and manganese levels tend to be lower and easier to remove due to the elevated dissolved oxygen (DO) levels. When rivers and streams are impounded, iron and manganese levels will increase.

The amount of iron and manganese that dissolves into the surface water depends on the character of the surrounding soil and the amount of plant life. Decomposition of organic matter (algae, leaves, and other plant material) in the lower sections of a reservoir may result in anaerobic conditions (i.e., near zero oxygen) under which iron and manganese compounds in those zones are converted to soluble compounds.

In much of North America, attempts to remove iron and manganese from impounded surface waters encounter a further challenge at least twice a year: just after the breakup of ice in the spring and just before water freezes in the fall or winter. Colder, denser water at the top of a reservoir sinks to the bottom, while warmer water at the bottom rises to the top, a phenomenon called a *water inversion* or *turnover*. The water that rises to the top brings with it soluble compounds of iron and manganese. As wind and wave action mixes oxygen from the air with the water, iron and manganese are gradually oxidized. The oxidized elements precipitate, or separate out of solution. These elements sink to the bottom again as solids and are brought back into solution where oxygen is absent and carbon dioxide has been formed. While some of the Mn^{2+} is very slowly oxidized in this way, the action of certain bacteria causes a much faster rate of oxidation. The scientific name for this process is microbially mediated oxidation. Microbially mediated Mn^{2+} oxidation can contribute significantly to manganese cycling in a reservoir.

Preparing for Inversion

An operator faces the challenge of learning how to read the signs of an inversion and how to adjust the treatment process to adapt to the increased levels of iron and manganese in the raw water coming into the treatment plant, usually through a combination of actions such as adding chemicals, shortening filter runs by backwashing more often, or other measures. If time and technology permit, measurements of DO should be taken, and a temperature profile should be developed to assist in identifying stratification. Records of operating data help operators know just when to prepare for increased levels of iron and manganese in the raw water. Because these occurrences do not necessarily happen at the same time every year, a strong case can be made for historical record keeping, which can illustrate patterns over time. Guided by records showing DO levels, stratification data, weather data, time lapses between events, and dates of events, the operator can respond proactively when levels of iron and manganese elevate.

It should be noted that gases are usually released as a part of the total water inversion occurrence, but by the time odors can be detected, iron and manganese have long since been released into the raw water.

Reducing Iron and Manganese in Feedwater

Because changes in the operation of the reservoir may reduce iron and manganese levels in the raw water supply, utilities should consider adjusting their reservoir management strategies or coordinating with the agency that operates the reservoir to determine the best actions to control these metals. During much of the year in temperate climates, the water nearest the reservoir surface is least likely to contain iron and manganese, but may contain other contaminants, such as algae. The inlet to the water treatment plant feedline should ideally move up and down, so water can always be taken from a level below surface plant growth but above the zone that lacks oxygen. Other iron and manganese reduction methods include installing an aeration system in the reservoir to increase oxygen levels or adding a mixing system to minimize stratification.

Some water treatment plants have intakes on streams or rivers located downstream of a reservoir. Unlike reservoirs, which generally have relatively low turbidity, streams and rivers can experience moderate to high turbidities during rain events (50 to 100 ntu) along with elevated iron and manganese associated with reservoir turnover. In some instances, high algae levels and taste-and-odor events occur simultaneously, adding to treatment challenges. Treatment and chemical feed systems that are flexible enough to address a wide range of water quality conditions should be considered.

HEALTH EFFECTS

Iron

Iron is an essential element in human nutrition. Estimates of the minimum daily requirement for iron depend on age, sex, physiological status, and iron bioavailability, and requirements range from about 10 to 50 mg/day. As a precaution against storage of excessive iron in the body, the Joint FAO[1]/WHO Expert Committee on Food Additives (JECFA 1983) established a provisional maximum tolerable daily intake (PMTDI) of 0.8 mg/kg of body weight. This applies to iron from all sources except iron oxides used as coloring agents, and iron supplements taken during pregnancy and lactation, or taken for specific clinical requirements. Allocation of 10 percent of this PMTDI to drinking water gives a value of about 2 mg/L, which does not present a hazard to health.

The taste and appearance of drinking water will usually be affected below this level, although iron concentrations of 1 to 3 mg/L can be acceptable for people drinking anaerobic well water. Iron (as Fe^{2+}) concentrations of 40 µg/liter can be detected by taste in distilled water. In a mineralized spring water with a total dissolved solids content of 500 mg/L, the taste threshold value was 0.12 mg/L. In well water, iron concentrations below 0.3 mg/L were characterized as unnoticeable, whereas levels of 0.3 to 3 mg/L were found acceptable (WHO 2003; Dahi[2] 1991).

Manganese

According to the USEPA, "Chronic (long-term) exposure to high levels of manganese by inhalation in humans may result in central nervous system (CNS) effects. Visual reaction time, hand steadiness, and eye-hand coordination were affected in chronically-exposed workers. A syndrome named *manganism* may result from chronic exposure to higher levels; manganism is characterized by feelings of weakness and lethargy, tremors,

---

[1] Food and Agriculture Organization of the United Nations

[2] Dahi, E., 1991. Personal communication.

a mask-like face, and psychological disturbances. Respiratory effects have also been noted in workers chronically exposed by inhalation. Impotence and loss of libido have been noted in male workers afflicted with manganism" (USEPA 2010).

Absorption. Limited data indicate that gastrointestinal absorption of manganese is low, averaging approximately 3 percent. The body has mechanisms that can usually control the total amount of manganese by increasing its elimination if excess levels are consumed. Because manganese and iron compete for gastrointestinal absorption, manganese uptake from the gut is likely to be increased in iron-deficient persons. Inhaled manganese may be absorbed in the lungs if the particles are small enough. Absorption through the skin is expected to be very low, less than 1 percent (Cooke 2014).

Short-term (acute) toxicity. Studies in animals and humans indicate that inorganic manganese has a very low acute toxicity by any route of exposure. However, acute inhalation exposure to high levels of manganese dust can cause an inflammatory response in the lung. In a human study, women took daily manganese supplements of 15 mg for 90 days. No toxic effects to the blood were seen (Cooke 2014).

Long-term (chronic) toxicity. Conclusive evidence that overexposure to manganese can result in adverse health effects has been observed in miners and steelworkers exposed by inhalation to manganese dusts. The range of effects seen is called manganism or referred to as *manganese-induced Parkinsonism*, because some of the symptoms are similar to those seen in cases of Parkinson's disease. Symptoms include muscle tremor, reduced motor skills, difficulty and slowing of walking, slurred speech, and sometimes psychiatric disturbances (Cooke 2014).

A few reports in the literature examine the effects of excess oral exposure of humans to manganese. One report describes symptoms of lethargy; increased muscle tone, spasm, and tremors; and mental abnormalities in persons drinking water contaminated by manganese from dry-cell batteries buried nearby. The actual length of time that people were drinking the water was not clear. The exposure time spanned a period of two to three weeks for some individuals and potentially much longer for others who developed symptoms. When the water was tested, it contained manganese at a concentration of 14.3 mg/L, which is well above the secondary standard set by USEPA. Those who studied this event believe it likely that other factors, possibly chemicals in the water besides manganese, contributed to the health effects (Cooke 2014).

The greatest exposure to manganese is usually from food. Adults consume between 0.7 and 10.9 mg/day in the diet, with even higher intakes being associated with vegetarian diets (Freeland-Graves et al. 1987; Greger 1999; Schroeder et al. 1966). Manganese intake from drinking water is normally substantially lower than intake from food. At the median drinking water level of 10 μg/L determined in the National Inorganic and Radionuclide Survey (NIRS), the intake of manganese from drinking water would be 20 μg/day for an adult, assuming a daily water intake of 2 liters. Exposure to manganese from air is generally several orders of magnitude less than that from the diet, depending on proximity to a manganese source (USEPA 2003). USEPA has also included manganese on its Contaminant Candidate List 4.

REGULATORY STANDARDS

According to the USEPA and WHO, iron and manganese in drinking water do not have adverse health consequences at the concentrations typically encountered in source waters. The USEPA and WHO have established a secondary MCL of 0.3 mg/L for iron and 0.05 mg/L for manganese. According to the USEPA, these secondary standards are established only as guidelines to assist public water systems in managing their drinking water for aesthetic considerations such as taste, color, and odor. An AWWA task group suggested limits of 0.05 mg/L for iron and 0.01 mg/L for manganese for an "ideal" water for public use (Bean 1962).

The Canadian drinking water quality guideline for iron is an aesthetic objective (AO) of less than or equal to 0.3 mg/L. The Canadian drinking water guideline for manganese is an AO of less than or equal to 0.05 mg/L.

It should be noted that the presence of manganese in drinking water will be objectionable to consumers if the manganese is deposited in water mains and causes water discoloration. Concentrations below 0.05 mg/L are usually acceptable to consumers, although this may vary with local circumstances.

According to Health Canada (1987), "The Recommended Daily Intake (RDI) of manganese for Canadians has yet to be established. In a recent comprehensive literature survey of studies of manganese metabolism in humans, it was concluded that previous estimates for a safe and adequate daily dietary allowance for manganese (2.5–5.0 mg/d) were too low, and a new range of 3.5–7.0 mg/d was recommended for adults."

Design Standards

In addition to the water quality regulations, several design standards describe needed treatment requirements. Several of these design standards are detailed in this handbook in the specific treatment sections.

Simultaneous Compliance

When addressing treatment for iron and manganese, the Safe Drinking Water Act should be considered in its entirety. Simultaneous compliance issues may include the following:

- Filter-top chlorination for manganese reduction, which can increase disinfection by-products
- Accumulation of radium in greensand filters
- The use of membranes to provide greater pathogen removal, which behaves differently from conventional filter sand with respect to manganese removal

To avoid simultaneous compliance issues, it is essential that all of the potential regulatory consequences be identified and that they be evaluated through studies, including jar and pilot testing.

Sustainability and Engineering Design

The production of safe drinking water traditionally has been associated with environmental and chemical science and engineering, and the focus historically has been on achieving treatment goals protective of public health and welfare. The objectives for the most part focused on achieving these goals in a cost-effective manner that considered the initial construction costs and long-term operational costs associated with energy, chemicals, labor, and residuals disposal. While up-front capital and long-term operating costs have played, and will continue to play, a major role in the decision-making process, the concept of *sustainable design and operation* has recently emerged and is being incorporated into projects today—much as environmental concerns were first considered in the 1960s, prompted by a then-new awareness of the impact the activities of mankind were imposing on the natural environment and fueled by the publishing of Rachel Carson's *Silent Spring*.

Most industries have now adopted policies promoting sustainability, and their executives often cite the importance of sustainability. For example, William Clay Ford Jr., Executive Chairman of the Board of Ford Motor Company, wrote (2010), "We continue to aggressively search for new ways, both big and small, to improve our economic and environmental sustainability. Often the actions we take accomplish both goals." This chapter defines how these newfound principles of sustainability can be defined in the context of, and incorporated into the design of, unit water treatment processes such as iron and manganese removal.

AN INDEFINITE DEFINITION

Sustainability has appeared as a global concept of concern over the past decade as a result of an acute awareness that the earth we live on, and many of the natural resources it provides, are, quite simply, finite in size and quantity. But what is sustainability? What does it mean, and more importantly in the context of this manual, how can such seemingly global concepts be applied to what is a specific process design?

The words *sustainability* and *sustainable* are used everywhere: brochures, ads, logos, business cards, magazines, and college curricula. If we asked our peers for a definition of the terms, we would likely uncover a plethora of perceived meanings, including some mention of energy, environment, quality of life, net zero, and the future. Some examples from a Hatch Mott MacDonald (HMM) survey of selected businesspeople include the following:

- "It means energy conservation and going green."
- "It means energy conservation and environmental acceptability."
- "Not sure, but it must be a good thing because everyone is talking about it."
- "It is impossible to attain, since everything we do affects the future."
- "It means whatever the user of the word wants it to mean."

On its home page for sustainability, the USEPA states (2014), "Sustainability is based on a simple principle: Everything that we need for our survival and well-being depends, either directly or indirectly, on our natural environment. Sustainability creates and maintains the conditions under which humans and nature can exist in productive harmony, that permit fulfilling the social, economic and other requirements of present and future generations. Sustainability is important to making sure that we have and will continue to have the water, materials, and resources to protect human health and our environment."

Many environmentalists and governments set the bar at unattainable heights with this vision: "Sustainable development seeks to meet the needs and aspirations of the present without compromising the ability to meet those of the future." This definition is from the well-known 1987 UN World Commission on Environment and Development report, known as the Brundtland Commission Report for the chair of the group, Gro Harlem Brundtland (United Nations 1987). However, most of what is produced now will in some way impact the future because every developed thing consumes natural resources, including land that future generations might need and current species inhabit.

The Center for Sustainable Engineering (CSE), a consortium of several prestigious universities, states, "Given the finite capacity of the earth, it is recognized that engineers of the future must be trained to make decisions in such a way that our environment is preserved, social justice is promoted, and the needs of all people are provided through the global economy" (CSE 2015). CSE member Carnegie Mellon further elaborates with the following conditional language: "Engineers of the future will be asked to use the earth's resources more efficiently and produce less waste, while at the same time satisfying an ever-increasing demand for goods and services. To prepare for such challenges, engineers will need to understand the impact of their decisions on built and natural systems, and must be adept at working closely with planners, decision makers, and the general public. Sustainable engineering emphasizes these and related issues" (Carnegie Mellon 2015). The Rochester Institute of Technology agrees: It considers sustainable engineering to be "the integration of social, environmental, and economic considerations into product, process, and energy systems design methods" (RIT 2015).

IMPORTANCE OF THE SUSTAINABLE APPROACH

Regardless of the multiple nuances in various organizations' attempts to define sustainability, a common thread exists in the mandate to consider (at least) three broad areas: economic, environmental, and social.

Perhaps, instead of adopting an almost impossible goal of avoiding the use of resources that would "impair future generations," it may be well to adopt a "zero net impact" goal. This would entail offsetting the unavoidable use of land, chemicals, fossil fuels, and the like by implementing social, environmental, and economic benefits that are equal to or greater than those sacrificed for the universal need to provide sources of drinking water that are protective of public health, as well as being economic (cost-effective) and compatible with the local environmental setting.

If a business or project cannot be run or built and operated economically, it cannot be sustained. The current economic climate, coupled with a program and project approach more accountable to the ratepayers, other stakeholders, and the public at large for the cost of services, has resulted in an increased emphasis on project capital and long-term costs. Sustainability, per se, should not be considered an added cost to projects and organizations. Instead, sustainability can be an approach to unleash innovation, growth, and efficiencies.

Once sustainability goals are defined for a project, the project should be designed to attain the desired goals at minimum cost. This is not to be confused with the idea that incorporating sustainability will lower the cost of a project. While cost-effectiveness is one of the pillars of sustainable design, the project stakeholders must define up front what measures are desired to be incorporated into the project to reflect the desired level of sustainability. Once that level and expected project outcomes are defined, the project designers' goal should be to attain those goals in as cost-effective a manner as possible.

These principles underscore the necessity for, and importance of, planning during the conceptual stages of a project. Sustainability measures are most successful when incorporated into the project right from the start. As the project progresses, technical and social momentum may hinder what could otherwise have been a fuller realization of sustainability goals.

Regardless of the entry point of the design engineer in the project timetable, the design team's progress toward sustainability goals must be continuously measured to provide accountability to the project stakeholders.

HOW IS SUSTAINABILITY MEASURED?

Societies throughout the world have recognized that the long-term environmental consequences of business practices need to be minimized, and concepts and tools are being developed to quantify the environmental impact and provide solutions to minimize these impacts. There are a number of scoring systems in use today for measuring and ranking the degree of sustainability in a given building or infrastructure. These systems are more applicable to the overall project than to individual unit processes. However, many of the principles contained in these systems can be directly applied to unit process design within a project. Two of the most prevalent systems being utilized in the United States are the Leadership in Energy and Environmental Design (LEED) system, and the Envision™ Sustainable Infrastructure Rating System (LEED 2014).

The LEED system is administered in the United States by the US Green Building Council (USGBC) and in Canada by the Canadian Green Building Council (CaGBC). LEED is a green building certification program that recognizes best-in-class building strategies and practices. To receive LEED certification, building projects satisfy prerequisites and earn points to achieve different levels of certification. The system has been used by a wide array of corporations, institutions, agencies, and residential and commercial developers. More than 30,000 projects in the US and more than 3,000 in Canada are now LEED certified.

The program is structured into seven key areas.

• Sustainable site development

• Water efficiency

• Energy efficiency

• Materials and resources

• Indoor environmental quality

• Innovation

• Adherence to regional priorities

Project designs are rated and scored using a point system. Points are tallied and compared to four rating levels: Platinum, Gold, Silver, and LEED Certified.

The Envision rating system is a tool that provides a holistic framework for evaluating and rating all types and sizes of infrastructure projects; it is administered by the Institute for Sustainable Infrastructure (ISI), located in Washington, DC. Comprising both public and private sector members, ISI was founded by three professional organizations: the American Society of Civil Engineers (ASCE), the American Council of Engineering Companies (ACEC), and the American Public Works Association (APWA). Envision itself is the product of a collaboration between ISI and the Zofnass Program for Sustainable Infrastructure at the Harvard University Graduate School of Design.

The Envision system evaluates projects against 55 assessment objectives organized into five categories: Quality of Life, Leadership, Resource Allocation, Natural World, and Climate & Risk. Points are assigned in each credit area depending on the planning and design parameters in that area. The points are tied to six "levels of achievement": Improved, Enhanced, Superior, Conserving, Restorative, and Innovative, each being defined in the program guidance. The evaluation results in an assessment of a project's overall "sustainability," and users of the program can apply for official ISI awards ranging from Bronze to Platinum. Table 2-1 presents the scoring system.

Table 2-1   Envision credit list

QUALITY OF LIFE – 13 Credits	
1. Purpose	QL1.1 Improve Community Quality of Life
	QL1.2 Stimulate Sustainable Growth & Development
	QL1.3 Develop Local Skills & Capabilities
2. Wellbeing	QL2.1 Enhance Public Health & Safety
	QL2.2 Minimize Noise and Vibration
	QL2.3 Minimize Light Pollution
	QL2.4 Improve Community Mobility & Access
	QL2.5 Encourage Alternative Modes of Transportation
	QL2.6 Improve Site Accessibility, Safety & Wayfinding
3. Community	QL3.1 Preserve Historic & Cultural Resources
	QL3.2 Preserve Views & Local Character
	QL3.3 Enhance Public Space
	QL0.0 Innovate or Exceed Credit Requirements
LEADERSHIP – 10 Credits	
1. Collaboration	LD1.1 Provide Effective Leadership & Commitment
	LD1.2 Establish a Sustainability Management System
	LD1.3 Foster Collaboration & Teamwork
	LD1.4 Provide for Stakeholder Involvement
2. Management	LD2.1 Pursue By-Product Synergy Opportunities
	LD2.2 Improve Infrastructure Integration
3. Planning	LD3.1 Plan for Long-Term Monitoring & Maintenance
	LD3.2 Address Conflicting Regulations & Policies
	LD3.3 Extend Useful Life
	LD0.0 Innovate or Exceed Credit Requirements
RESOURCE ALLOCATION – 14 Credits	
1. Materials	RA1.1 Reduce Net Embodied Energy
	RA1.2 Support Sustainable Procurement Practices
	RA1.3 Use Recycled Materials
	RA1.4 Use Regional Materials
	RA1.5 Divert Waste from Landfills
	RA1.6 Reduce Excavated Materials Taken Off-Site
	RA1.7 Provide for Deconstruction & Recycling

(continued)

Table 2-1   Envision credit list (continued)

2. Energy	RA2.1 Reduce Energy Consumption
	RA2.2 Use Renewable Energy
	RA2.3 Commission & Monitor Energy Systems
3. Water	RA3.1 Protect Fresh Water Availability
	RA3.2 Reduce Potable Water Consumption
	RA3.3 Monitor Water Systems
	RA0.0 Innovate or Exceed Credit Requirements

NATURAL WORLD – 15 Credits

1. Siting	NW1.1 Preserve Prime Habitat
	NW1.2 Protect Wetlands & Surface Water
	NW1.3 Preserve Prime Farmland
	NW1.4 Avoid Adverse Geology
	NW1.5 Preserve Floodplain Functions
	NW1.6 Avoid Unsuitable Development on Steep Slopes
	NW1.7 Preserve Greenfields
2. Land and Water	NW2.1 Manage Stormwater
	NW2.2 Reduce Pesticide & Fertilizer Impacts
	NW2.3 Prevent Surface and Groundwater Contamination
3. Biodiversity	NW3.1 Preserve Species Biodiversity
	NW3.2 Control Invasive Species
	NW3.3 Restore Disturbed Soils
	NW3.4 Maintain Wetland & Surface Water Functions
	NW0.0 Innovate or Exceed Credit Requirements

CLIMATE & RISK – 8 Credits

1. Emissions	CR1.1 Reduce Greenhouse Gas Emissions
	CR1.2 Reduce Air Pollutant Emissions
2. Resilience	CR2.1 Assess Climate Threat
	CR2.2 Avoid Traps & Vulnerabilities
	CR2.3 Prepare for Long-Term Adaptability
	CR2.4 Prepare for Short-Term Hazards
	CR2.5 Manage Heat Island Effects
	CR0.0 Innovate or Exceed Credit Requirements

CORE SUSTAINABILITY FACTORS

The LEED and Envision rating systems provide proven methods of approaching a project that may be adopted for an iron and manganese removal facility. These approaches can be organized into two main categories: overall project approaches and specific design approaches inherent to the iron and manganese removal processes.

Sustainable outcomes to consider and factor heavily in the forming of project goals include the following:

- Minimizing the use of nonrenewable resources
- Fostering collaboration and teamwork among all stakeholders
- Using recycled materials and local materials (spurring the local economy and minimizing construction material transportation costs) and labor (developing local skills and capabilities) for cost-effective construction
- Minimizing the overall carbon footprint of the facility (it has been shown that an overall reduction in the carbon footprint of a project will reduce life-cycle costs)
- Maximizing energy efficiency and use of renewable energy sources
- Reducing waste produced during construction as well as throughout the life of the project
- Engaging the community in the siting and design of the project
- Providing enhancements to the local community (e.g., trails and park settings in facility buffers)
- Preserving habitats, greenfields, and surface water resources beyond those required by local, state, and federal regulation
- Thoroughly assessing climate threats to the facility to provide built-in defenses against facility damage and service interruptions
- Improving local community mobility, access, and safety
- Using alternative materials with life cycles that align with the project lifetime
- Using brownfields or other recyclable commercial/industrial sites, which often provide isolated locations and may otherwise remain unused

SUSTAINABLE IRON AND MANGANESE REMOVAL

The main components of an iron and manganese removal system are

- Water supply source (well, reservoir, river/stream)
- Facilities housing the system components
- Treatment system (chemical, filtration)
- Residuals management system

These project components may be housed together or separately, and the size and type of building(s) will dictate the applicability of the sustainability measures employed (e.g., the LEED system). Building location, orientation, materials of construction, lighting systems, and overall aesthetics are key features amenable to the incorporation of sustainable design approaches. For example, sustainability goals for the project buildings may include the following:

- Taking maximum advantage of solar energy and building orientation to minimize heating and cooling costs
- Using materials of construction that blend well into the natural environment of the site setting
- Using recycled materials in the building construction
- Incorporating green roofs into the building designs
- Installing porous pavement and other stormwater runoff mitigation measures
- Using natural lighting during the day and limited/focused lighting at night to minimize light pollution
- Using solar or wind energy to supply the electrical needs of the building (and possibly the equipment)
- Considering the use of readily reused or recycled materials for the buildings for a more sustainable approach to end-of-life demolition

An example of a building incorporating sustainable features is shown in Figure 2-1.

Sustainable activities for the water supply source may include the following:

- Using aeration or mixing systems in a reservoir to minimize the intake of iron and manganese
- Evaluating solar power for the mixing systems
- Replacing pumps with higher-efficiency units

- Using materials with a lower carbon footprint for construction of the well and pumping systems

As for the treatment process, the selection of the iron and manganese removal process to be utilized is increasingly affected by a key sustainability concern: the scarcity of conventional manganese greensand that uses glauconite as the base media. The conversion of glauconite to greensand requires extensive and complex processing and the disposal of waste products. More sustainable filter media are now available that have less environmental impact.

Other sustainable components of the treatment system may include the following:

- Using renewable products for filter media
- Using materials with a lower carbon footprint for construction of the treatment equipment
- Generating sodium hypochlorite on site as a substitute for bulk sodium hypochlorite
- Optimizing operational energy usage through measures such as minimizing head loss across the entire system and improving the backwash system to effectively clean the media while reducing energy usage and residuals production
- Conducting energy audits during the replacement or expansion of existing facilities, resulting in immediate savings by efforts such as replacing inefficient motors

Sustainable components of the residuals system may include the following:

- Reducing offsite residuals disposal by treating the residuals and recycling the decant
- Developing alternative uses for the residuals to reduce disposal to a publicly owned treatment works or landfill

LOOKING AHEAD

Achieving sustainable iron and manganese removal projects need not be difficult. Identifying sustainability goals at the conceptual stage of a project is a key element leading to project success. Although potentially confusing and daunting at the start, a straightforward stepwise approach to sustainability based on a fundamental understanding of proven concepts, and a logical approach to the various components of the treatment system, will ensure that the project stakeholders' goals will be achieved.

Figure 2-1   Sustainable design for chlorination station. This design for a chlorination station incorporates a number of sustainable features, including green walls and roofs, permeable pavers, and the replacement of mowed lawns with wildflower meadows.
Source: Hatch Mott MacDonald

The remaining chapters of this handbook discuss each of the components that are a part of the iron and manganese removal treatment process. The use of sustainable technologies and strategies is possible with each component.

3

Basic Chemistry

An understanding of the basic concepts associated with iron and manganese is important because the chemistry affects the occurrence and forms of iron and manganese as well as the type of treatment that can be used for controlling and removing these elements. For example, chlorine is often used for oxidation of manganese. However, some surface water facilities have changed to permanganate-based products because of elevated disinfection by-products. These alternate oxidants may result in unanticipated issues such as filter clogging or changes in treated water quality. A detailed description of the oxidation processes is presented in chapter 5.

Simple soluble reduced iron is ferrous iron (Fe^{2+}), and simple soluble reduced manganese is manganous manganese (Mn^{2+}). When oxidized, the iron changes to ferric iron (Fe^{3+}), and manganese often (but not always) changes to the insoluble Mn^{3+} or Mn^{4+} forms. Both iron and manganese can form combinations with many other elements and compounds. A treatment plant operator need not memorize the periodic table of the elements, understand atomic structure, or know how to calculate molecular weight, but it is vitally important for an operator to understand some basic changes that take place when certain chemicals come into contact with certain other elements and compounds. For example, the presence of organic carbons, hydrogen sulfide, or ammonia often indicates a potential for interference with iron and manganese removal processes. Thus, an operator should have a basic knowledge of organic carbons and ion exchange principles.

CHEMICAL SYMBOLS, COMMON FORMULAS

A chemical element may be defined as a chemical substance that cannot be broken down into simpler substances by ordinary chemical change. Water, a chemical compound, can break down further into hydrogen and oxygen (which are chemical elements). Hydrogen and oxygen cannot break down into still simpler substances without altering their basic identities.

The smallest part of an element is an atom; it has all the chemical properties of the element in which it exists. Smashing an atom produces new elements of lighter weight and releases a tremendous amount of energy.

In the language of chemistry, each element is represented by a symbol. A combination of symbols to characterize a chemical compound is called the formula for that compound. A formula gives all the elements in the chemical compound and the number of atoms of each:

- Symbol for hydrogen: H
- Symbol for oxygen: O
- Formula for water: H_2O; the compound water has two atoms of hydrogen combined with one atom of oxygen. Note that single atoms are not numbered.

IONIZATION

Many elements combine with each other on contact. This change can happen slowly, rapidly, or explosively, often under the influence of heat, moisture, or catalytic agents (substances that promote reactions but do not themselves undergo any change during the reactions). Many chemical reactions take place in the presence of water, because water can split some molecules into positively or negatively charged atoms called ions. This process is known as *ionization*. A molecule that produces ions when dissolved in water is said to ionize.

Example: $NaCl \rightarrow Na^+ + Cl^-$. The electrically charged atoms are ions.

A positively charged ion is called a *cation* and a negatively charged ion is referred to as an *anion*. The charges must be equal in strength and opposite in sign. Otherwise the solution itself would have a charge, which is impossible.

When a molecule ionizes in water, the charges must equalize even when the molecule combines more atoms of one kind than another. For example, $FeCl_3$ (ferric chloride) ionizes like this.

$$FeCl_3 \rightarrow Fe^{+3} + 3Cl^-$$

Some molecules ionize to yield groups of atoms, consisting of positively charged cations and negatively charged anions. For example, Na_2SO_4 (sodium sulfate) ionizes like this: $Na_2SO_4 \rightarrow 2Na^+ + (SO_4)^{-2}$, leaving two sodium cations and one sulfate anion. Combinations of atoms in a unit [such as $(SO_4)^{-2}$] are called *radicals*.

pH VALUE

Water ionizes to a slight degree, producing both hydrogen (H^+) and hydroxyl [$(OH)^-$] ions, as follows:

$$H_2O \leftrightharpoons H^+ + (OH)^-$$

Water may be described as both an acid and a base (alkaline), because it produces both hydrogen and hydroxyl ions.

Because these ions are present in identical concentrations, however, pure water is neutral and therefore has no charge. But when water ionizes, the extent of the ionization depends on the relative concentration of hydrogen and hydroxyl ions. A complicated mathematical formula for this phenomenon produces Table 3-1, which describes the relative acidity or alkalinity of any water sample. The pH scale goes from 0 to 14, neutral water having a pH of 7.

Careful examination of Table 3-1 discloses several important facts.

1. The highest hydrogen ion concentration corresponds to the lowest pH.
2. The lowest hydrogen ion concentration corresponds to the highest pH.
3. Neutrality at pH 7 defines the midpoint in the scale. pH values lower than 7 represent hydrogen ion concentration above neutrality (acidity), and pH values higher than 7 represent hydrogen ion concentration below neutrality (alkalinity).
4. In practice, pH values can fall between whole numbers. For example, one water's pH could be 6.1, another water's pH 6.2, and still another water's pH 9.5. Because pH is logarithmic, fractional differences in pH can have a significant impact on water chemistry.

This variation shows once again the importance of ionization, the major role played in chemistry by hydrogen and hydroxyl ions, and the need for an operator to know just how acidic or basic (alkaline) a water is.

Table 3-1   Ionization and pH value

A Grams of H+ ions per liter	B Reciprocal of A	Log B	C (pH)
1.0	1.0	0	
0.1	10	1	
0.01	100	2	
0.001	1,000	3	0–6: Acid
0.0001	10,000	4	
0.00001	100,000	5	
0.000001	1,000,000	6	
0.0000001	10,000,000	7	7: Neutral
0.00000001	100,000,000	8	
0.000000001	1,000,000,000	9	
0.0000000001	10,000,000,000	10	
0.00000000001	100,000,000,000	11	8–14: Alkaline
0.000000000001	1,000,000,000,000	12	
0.0000000000001	10,000,000,000,000	13	
0.00000000000001	100,000,000,000,000	14	

The acidity of a water sample is measured on a pH scale. This scale ranges from 0 (maximum acidity) to 14 (maximum alkalinity). The middle of the scale, 7, represents the neutral point. The acidity increases from neutral toward 0.

Because the scale is logarithmic, a difference of one unit represents a tenfold change. For example, the acidity of a sample with a pH of 5 is 10 times greater than that of a sample with a pH of 6. A difference of two units, from 6 to 4, would mean that the acidity is 100 times greater, and so on.

Normal rain has a pH of 5.6, slightly acidic due to the carbon dioxide absorbed from the earth's atmosphere by the rain.

Figure 3-1 shows the range of the pH scale by identifying values for some common substances. As noted in the figure, "a difference of one pH unit represents a tenfold change." For example, lye with a pH of 13 is 10 times more alkaline than ammonia with a pH of 12.

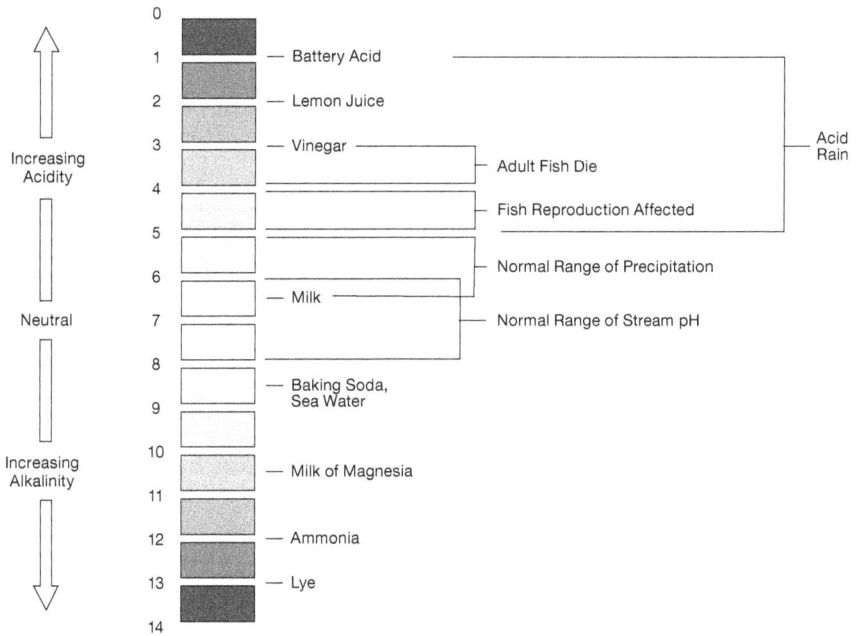

Figure 3-1   pH scale

Many chemical reactions are either accelerated or slowed by pH adjustments. For example, manganese (Mn^{+2}) oxidizes many times more slowly at pH 7 than it does at pH 8.5.

No single finished water pH applies as a standard for all treatment plants. Finished water pH targets should be set by distribution system and consumer issues, including corrosion considerations. Water with pH 8 may be an answer to a corrosion problem in one plant, while water with pH 7 will be the answer in another plant.

CHEMICAL FORMS OF IRON AND MANGANESE

Iron in its simplest soluble form, the form most often encountered in raw well water or raw impounded water, is ferrous iron. Manganese in its simplest soluble form is manganous manganese. Both have a valence of 2 (i.e., Fe^{2+} and Mn^{2+}).

Understanding Valence Numbers

Valence, touched upon earlier in the chapter, is the number of electrons that an atom or radical can lose, gain, or share with other atoms or radicals. Valence has the same meaning as combining power. One may say it is the relative worth of an atom of an element in combining with the atoms of other elements to form compounds. A positive or negative whole number ranging from 0 to 7 expresses the valence of an atom or radical.

If an atom loses an electron from its outermost shell, it is said to have a valence of 1. Similarly, if an atom loses two electrons from its outermost shell, it is said to have a valence of 2. If the outermost shell gains electrons, the atom has a negative valence. An atom that gains one electron has a valence of -1. One that gains two electrons has a valence of -2, and so on.

The manganese in $MnO_2(s)$, such as pyrolusite and the artificial coating on manganese greensand, has a valence of 4, expressed as Mn^{4+}. The manganese in potassium permanganate ($KMnO_4$) has a 7 valence, expressed as Mn^{7+}. In forming the chemical compound $KMnO_4$, the manganese atom lost (or gave up) seven electrons.

An element's valence number states the number of electrons of the element associated with formation of a particular compound. Because this number may differ for different compounds of an element, it follows that any element may have more than one valence number.

For example, iron combines with oxygen to form two different compounds: FeO, a black oxide, and Fe_2O_3, a red oxide composed of two atoms of iron and three atoms of oxygen. The red oxide is commonly referred to as rust. The black oxide is frequently observed in water plants, particularly underneath the scale of corroded pipe. Iron also forms many compounds with other elements in which it may have a valence number of 2 or 3.

The reasons for some facts of chemical life are unknown. Manganese can have valence numbers 2, 3, 4, 6, and 7.

- Mn^0 is manganese metal.
- Mn^{2+}, pale pink in color and soluble, is the manganous ion.
- Mn^{3+}, red-violet in color and insoluble, is the manganic ion.
- Mn^{4+}, brown-black in color, is the valence of the manganese in solid pyrolusite, $MnO_2(s)$.
- Mn^{6+}, dark bottle-green in color, is the manganate ion in MnO_4.
- Mn^{7+}, intense purple in color, is the permanganate ion in MnO_4 (as in $KMnO_4$).

ORGANIC CHEMISTRY

This field is the specific branch of chemistry concerned with compounds of carbon. Carbon is unique among all other elements, not only because of its great reactivity, but also because of its ability to combine with itself in various ways to form a very large number of compounds. Many thousands of carbon compounds are known, and more are being discovered. The possibilities seem endless, especially as chemists design synthetic compounds for special purposes. All plant and animal tissue is composed of numerous carbon atoms in combination chiefly with hydrogen and oxygen and less commonly with nitrogen, phosphorus, sulfur, and metals.

Not all carbon compounds, by any means, are good for people, though. For example, some by-products resulting from combinations of chlorine and organic carbons are known as trihalomethanes (THMs). Under laboratory conditions, THMs have caused cancer in animals, and are therefore known as *carcinogens*. Obviously, every water treatment plant operator should recognize the dangers of producing chemical by-products harmful to humans. For example, the old anesthetic chloroform is trichloromethane ($CHCl_3$), one member of the THM grouping of undesirable by-products.

Carbon has a valence of +4. To understand this characteristic, picture the carbon atom as having four hooks. Likewise, a hydrogen atom has one hook, oxygen has two hooks, nitrogen has three hooks, ferrous iron has two hooks, and ferric iron has three hooks. The concept gets more complicated with some elements, such as manganese, which has two-, three-, four-, six-, and seven-hook variations. The "hooks" are called *bonds* or *valence*.

Organic Complexing of Iron and Manganese

A treatment plant operator is expected to find a practical, safe, effective, affordable way of meeting iron and manganese removal objectives. At the same time, the operator should have some knowledge of organic chemistry to allow identification of certain characteristics of the raw water being treated, without expending great amounts of time and money conducting exhaustive scientific tests. At the least, an operator needs to know the information presented in this section.

Recurring phenomena may display almost identical early warning signs of iron and manganese complexing, indicating at the very least the presence of organic compounds. Such a condition suggests a need to depart from preoxidation/sedimentation/filtration processes. Removing iron and manganese (especially manganese) below targeted levels typically involves the following difficulties:

1. Well water is being drawn from shallow wells adjacent to flowing water, lakes, or sloughs.
2. A level of organic carbon over 2 mg/L is identified.
3. Some level of ammonia is quantified.
4. Some level of hydrogen sulfide is detected.

If the preoxidation/sedimentation/filtration process cannot remove iron and manganese below target levels in a situation marked by any one or a combination of these four factors, the iron and manganese might be removed to acceptable levels by treating them as organically complexed.

Organic complexing is the process through which such elements as iron and manganese become part of a carbon compound. Complexation is a chemical reaction in which certain chemicals "tie up" other chemicals, particularly metal ions, so that the chemicals no longer react.

The most troublesome of organic compounds in iron and manganese removal processes are the organic acids, which contain one or more carboxylic groups (COOH). In this carboxyl group, the hydrogen (H) is "active," i.e., it will ionize slightly to give acidic H^+ ions. More importantly, the active hydrogen will exchange for ions such as calcium, magnesium, sodium, iron, manganese, nickel, etc., exactly as occurs in zeolite softening. The common, naturally occurring organics encountered in water result from decomposition of organic materials (e.g., plants, leaves, and algae). A common organic compound is humic acid, which has the approximate formula

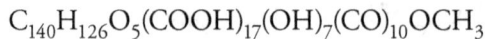

$$C_{140}H_{126}O_5(COOH)_{17}(OH)_7(CO)_{10}OCH_3$$

Chemists have defined no exact formula for this acid. A molecule of humic acid contains about 17 carboxylic groups, so it shares some characteristics with a water-soluble, ion-exchange resin. Some ion-exchange resins (zeolites) are water insoluble and contain thousands of carboxylic sites. Called weak acid cation exchange resins, they are used widely in the water industry.

Anionic polymers also contain the carboxyl as their active sites. As H ionizes to H^+, the rest of the carboxyl group has a negative charge.

This discussion relates to iron and manganese removal because hydrogen of the carboxyl group will exchange with Fe^{+2} and Mn^{+2} ions, each of these soluble, two-valent ions exchanging with two hydrogens. This process is commonly described as *organic binding* or *complexing*.

An attempt to oxidize the bound Fe^{+2} or Mn^{+2} may or may not succeed. If oxidation occurs, creating insoluble ferric iron (Fe^{+3}) or changing Mn^{+2} to insoluble Mn^{+3} or Mn^{+4}, the iron or manganese may still be held by the ion-exchange bonding. In such a case, the process may yield complexed, oxidized Fe/Mn forming a tiny organics/Fe/Mn clump or colloid, which carries an overall negative charge and is very difficult to remove, as it does not settle and passes through filters. A nonionic or anionic polymer may help to agglomerate colloids to form units large enough to settle out or become trapped in filters. Organic complexing can result from ion exchange, although not all chemical binding works this way.

Attempts to precisely identify the organics present in raw water require extremely difficult and costly methods seldom justifiable as a routine control measure. Instead, most operators monitor total organic carbon (TOC) as an indicator to warn of a potential for organic complexing. TOC levels greater than 2.0 mg/L to 2.5 mg/L indicate a potential removal problem using the preoxidation/direct filtration process.

It may also be possible to establish a correlation between UV_{254} and TOC. UV_{254} is much less expensive to measure than TOC, several types of UV analyzers are available, and UV can be measured using on-line instruments.

Jar testing is recommended to assess the effectiveness of any oxidant/coagulation regime. Iron and manganese that are organically complexed can usually be oxidized using chlorine or $KMnO_4$, given the appropriate dosage, pH, and detention time. Organically complexed iron can sometimes be oxidized by oxygen from simple aeration. In this state, its size is usually very fine or colloidal, so it is too small to be removed by a granular media filter. One removal option is to neutralize the surface charge using coagulants such as alum, iron salts, poly aluminum compounds, or cationic polymers, followed by settling and/or direct filtration.

Oxidized iron and manganese species are not organically complexed, and their compounds can remain colloidal following oxidation. Again, the removal method described in the previous paragraph is one option. In some cases, however, the raw water contains elements and compounds that interfere with the action of certain coagulants.

Going Beyond the Traditional

This chapter provides only an overview of the behavior of iron and manganese. If the operator is unable to remove excessive iron and manganese, the reason could be interference from other compounds. As water becomes degraded over time, remedies that worked in the past may no longer be effective.

It may be necessary to go beyond traditional approaches and to experiment in a logical fashion in order to find an effective solution. It may be necessary to try a number of approaches that do not work before arriving at the answer.

Overview of Treatment Technologies

Several treatment methods may be used to remove iron and manganese from drinking water supplies. This chapter provides an overview of treatment options that should be considered for iron and manganese removal and includes guidance regarding selection of treatment methods for a particular application.

Over the past decade or so, advances in treatment technologies have allowed systems to be constructed in less space, produce fewer residuals, and treat multiple contaminants.

This chapter addresses two key questions:

1. Why consider new technologies?
2. When should traditional technologies be used?

A few years ago, a utility discovered that a well with elevated iron (10 mg/L) and manganese (0.5 mg/L) also showed evidence of surface water contamination. Traditional treatment methods would have included the use of clarification, oxidation, coagulation, and filtration. Instead, the utility was able to use an ultrafiltration (UF) membrane to accomplish the removal of the iron, manganese, and pathogens. No clarification or coagulation was needed. The use of membranes was—and in some cases still is—considered to be an innovative treatment technology. The use of membranes resulted in significant savings for the owner. Figure 4-1 describes the traditional and newer technologies that can be used to remove iron and manganese.

SOURCE OF SUPPLY

- Well
- Reservoir
- Creek/River

OXIDATION

- Aeration
- Chlorine (gas chlorine, sodium hypochlorite, calcium hypochlorite, on-site generation of sodium hypochlorite)
- Potassium Permanganate
- Sodium Permanganate
- Chlorine Dioxide
- Ozone

CLARIFICATION

- Conventional Sedimentation
- Solids Contact Clarification
- Plate Settlers
- Tube Settlers
- Ballasted Flocculation

RESIDUALS

- Direct Discharge to Sewer
- Equalization and Sewer
- Lagoons and Sand Drying Beds
- Mechanical Dewatering

FILTRATION

- Dual Media
- Greensand
- Ion Exchange
- MnO_2 Coated Media
- MnO_2 Ore
- Hollow-Fiber Membranes
- Spiral Membranes
- Ceramic Membranes
- Biological Filtration

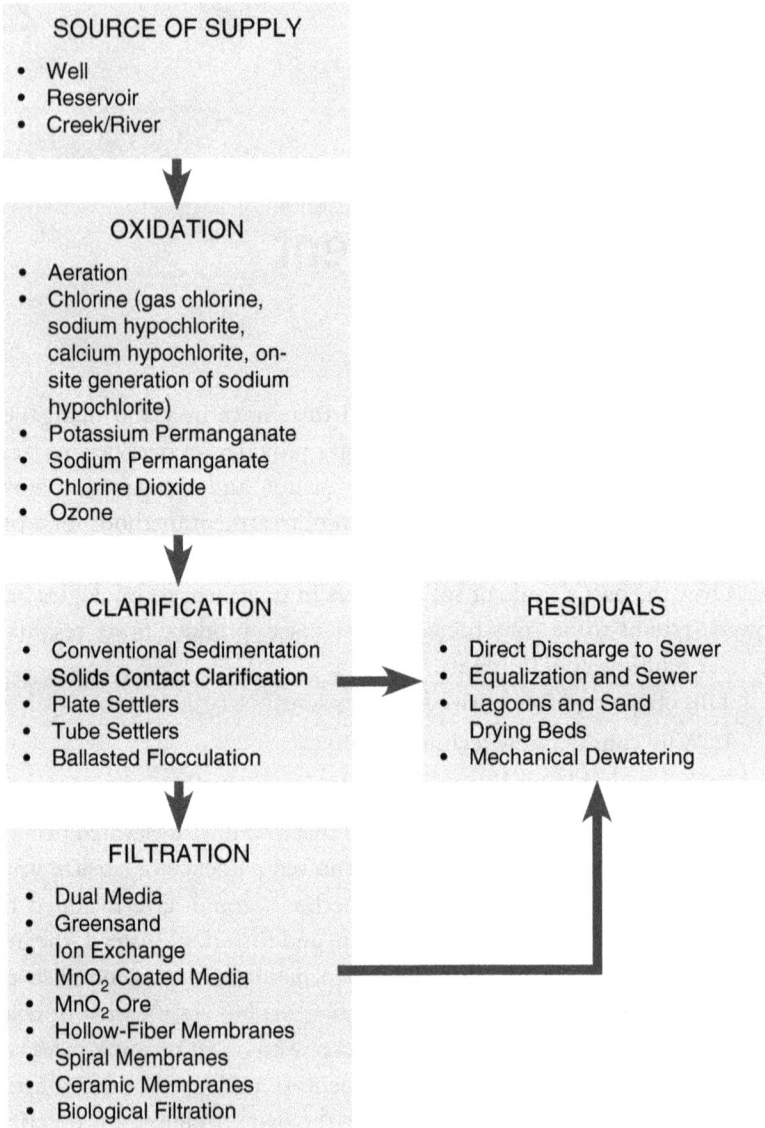

Figure 4-1   Technologies for iron and manganese removal
Source: Hatch Mott MacDonald

OVERVIEW OF TECHNOLOGIES

The treatment for iron and manganese typically involves multiple unit processes. The answer to the question "When should traditional technologies be used?" can be found in the properties of iron and manganese that are encountered in source waters.

As discussed in chapter 1, iron and manganese are often found in water supplies in their dissolved state. Traditional technologies involve converting these dissolved forms to a particulate form that can be clarified and filtered. The conversion from dissolved to particulate forms requires the use of an oxidant. Traditional forms of oxidation are well understood and can be successfully employed.

Conventional sedimentation and filtration technologies are well suited for the removal of particulate forms of iron and manganese. Traditional technologies are typically simple to operate and generally do not employ the use of proprietary treatment technologies.

Innovative technologies may be beneficial for small sites or in situations that have multiple contaminants. Nontraditional technologies may warrant consideration for systems with the following conditions:

- High concentrations of radionuclides
- Wells under the influence of surface water
- Wells with high iron/manganese and limited space
- Sites where disposal of residuals is problematic
- Sites where very high backwash volumes would impact water production

The nontraditional treatment methods for the removal of iron and manganese include the following:

- Ballasted flocculation
- Membrane filtration
- Biological filtration

When selecting and designing a process for iron and manganese treatment, the quality of the source water should be analyzed comprehensively, because some contaminants can affect the operation of the system. For example, radiological contaminants may be removed by some types of filter media, but the media may become radioactive, resulting in safety concerns and increased cost for media disposal. In most cases, separate treatment for radiological contaminants may result in a lower life-cycle cost.

SEQUESTRATION

Sequestration is a form of treatment in which a chemical, known as a *sequestrant*, is added to groundwater. The chemical forms a bond with iron and manganese ions, allowing them to remain in solution. Sequestration for drinking water treatment of iron and manganese is generally limited to sources where the iron is less than 0.6 mg/L and the manganese is less than 0.1 mg/L. Sequestration of source water concentrations above these values may result in aesthetic issues in the distribution system and is generally not allowed by regulators.

OXIDATION, CLARIFICATION, FILTRATION

Most iron and manganese removal treatment processes incorporate oxidation to convert the dissolved forms of the metals to a solid, followed by a filtration process. When concentrations in the source water are above 8 to 10 mg/L combined iron and manganese, a clarification step is typically required prior to filtration. A combined iron and manganese concentration of 8 mg/L will generally result in a filter run time of less than 24 hours for sand/anthracite filters as well as greensand-type filters.

The clarification process reduces the amount of solids that must be removed by the filters, which results in longer filter run times before backwashing is required. Clarification may be achieved using any number of technologies, such as conventional sedimentation, plate or tube settlers, solids-contact clarifiers, dissolved air flotation, or ballasted flocculation. The type of filtration used in this treatment approach may be a conventional dual-media anthracite-and-sand filter, a mono-medium sand or anthracite filter, glauconite, manganese dioxide–coated sand media, or manganese dioxide ore.

OXIDATION, MANGANESE DIOXIDE–COATED MEDIA FILTRATION

Filtration using manganese dioxide–coated and ore-type catalytic media, such as manganese greensand, has historically been used for iron and manganese removal. In recent years advances have been made in the types of manganese dioxide media, allowing for higher loading rates and reducing backwash requirements.

Several different types of proprietary manganese dioxide–coated media filtration systems are available for iron and manganese removal. Although the mechanics vary, the basic treatment provided is oxidation of iron and manganese with the addition of chlorine or potassium permanganate followed by filtration of precipitates. Manganese dioxide coating on the filter media acts as a catalyst for the oxidation and reduction of iron and manganese.

Some of the proprietary media on the market include the following:

- Ferrosand® manganese greensand produced by Hungerford & Terry
- GreensandPlus™ synthetic manganese greensand produced by Hungerford & Terry
- LayneOx™ produced by Layne Christensen Company
- ANTHRA/SAND™ produced by WesTech

These media can be classified into two groups: manganese dioxide–coated sand and manganese dioxide ore. All these media are similar in that they use a combination of chemical oxidation and catalytic media to remove iron and manganese in water. Different proprietary media have different specific mechanics, loading rates, backwashing requirements, and removal efficiencies.

Design of the manganese dioxide–coated media systems is discussed in further detail in chapter 7.

OXIDATION, MEMBRANE FILTRATION

Membranes are often used for removal of turbidity and pathogens from surface water and groundwater under the direct influence (GUDI) of surface water. Membranes are also used for iron and manganese removal. Membrane treatment is often a viable option for GUDI wells that require treatment for iron and manganese. Membranes may be considered as a treatment option when the source water is non-GUDI groundwater, but typically are not the most cost-effective treatment solution in this situation.

Membrane systems operate by straining out particles that are larger than the pore size of the membrane. Microfilters are generally classified as having openings between 0.05 and 1.0 microns, whereas ultrafilters are generally classified as having openings between 0.005 and 0.05 microns. Dissolved iron and manganese converted to particulate form via conventional oxidation can be subsequently strained out on the membrane.

Several manufacturers of micro/ultrafilters have experience in treating water for iron and manganese. Each system uses a unique membrane and operates slightly differently in terms of backwashing and cleaning. Pressurized and immersed membrane systems are available.

BIOLOGICAL FILTRATION

Hydraulically, the biological filtration process operates similarly to a pressure filter in that raw water is pumped through a pressure vessel containing a granular media. However, unlike most other pressure filtration systems, which rely on the formation of a chemical precipitate and subsequent filtration, biological processes do not require any chemical oxidants.

Instead, conditions are established in the pressure vessel that foster the growth of bacteria. These bacteria oxidize the iron and manganese in the raw water, which is then retained within the filter in the form of dense precipitates. These precipitates are more compact than the amorphous precipitates formed during chemical oxidizing processes. Therefore the biological filter has a higher iron and manganese retention capacity (up to five times higher). The increased metal retention capacity allows the system to achieve long filter run times. Air is continuously injected into the raw water to provide the proper growth environment for the bacteria.

It is important to note that the required environmental conditions for biological iron removal are different than those for biological manganese removal. Therefore, where both iron and manganese are present, two stages of biological filtration are required: one for biological removal of iron and one for biological removal of manganese.

Biological filtration processes for removal of iron and manganese are proprietary patented systems manufactured by Infilco Degremont, Inc., marketed under the names Ferazur® and Mangazur®. Design of biological filtration systems is discussed in further detail in chapter 7.

RESIDUALS

All iron and manganese removal processes generate residuals. The requirements for residuals treatment will differ at each site depending on land available, the viability of disposal of waste to a sanitary sewer, the feasibility of recycling to the head of the plant, and other factors. The residuals generated by the treatment process and the options for treatment should be considered when selecting an iron and manganese removal process.

Some residuals treatment processes that may be considered include the following:

- Direct sewer discharge
- Equalization followed by sewer discharge
- Batch settling with decant recycle and solids discharge to sewer

- Equalization followed by treatment using plate settlers with decant recycle
- Lagoons
- Mechanical dewatering

Chapter 8 provides a discussion of residuals handling.

EVALUATION AND SELECTION CRITERIA FOR OPTIMUM TREATMENT

The process of evaluating and selecting a treatment option involves quantifying the iron and manganese concentrations in the source water, identifying other water parameters that require or impact treatment, and establishing the finished water quality goals.

Understanding the project constraints, such as site conditions, sewer availability, and regulatory acceptance, will inform the selection process. Initial technologies should be screened for their potential feasibility and a subset of unit processes established for further evaluation. The evaluation generally consists of the preparation of concept-level designs for each option, including the development of a flow schematic, building floor plan, section, and site plan. These drawings are then used to develop construction, operating, and life-cycle costs.

In addition to the cost estimates, other factors should be considered. The technical and financial feasibility of a treatment option should be the primary criteria, but qualitative issues should also be weighed where more than one viable treatment option exists. Some qualitative considerations may include sustainability, operational flexibility, maintenance requirements, ease of integration into an existing facility, complexity of the process, track record of the proposed technology on similar applications, and location of service technicians.

The full list of qualitative issues that should be considered will vary from site to site and client to client. An example evaluation methodology (qualitative ranking matrix) is presented in Table 4-1. In order to develop a collaborative input/output for the qualitative analysis, the project team should consider completing the ranking matrix with the input of the owner and operations staff to establish the relative weight and rating components.

Table 4-1   Qualitative ranking matrix for determining suitability of an iron and manganese removal system

Criteria	Relative Weight (0–5)	Rating (1–5)	Weighted Rating
Overall system criteria			
Confidence in technology for iron and manganese removal			
Confidence of the technology for crenoform removal			
Equipment complexity—number of pumps and compressors			
Location of service technicians			
Operating costs			
Residuals			
Residuals generation as percent of overall plant output			
Impact of residuals recycling on plant performance			
Level of treatment required for residuals			
Ability to send residuals to sanitary sewer			
Operational criteria			
Level of operator attention required			
Ease of maintenance			
Experience in treatment chemicals used			
Ease of startup and shutdown of system			
Layout-related criteria			
Ease of integration with existing facilities			
Layout flexibility—restrictions on equipment location			
Total Weighted Rating			

TREATMENT SYSTEM DESIGN ELEMENTS AND WORK FLOW

The design of an iron and manganese treatment system begins with the process engineer defining the water quality and flow rate requirements of the system. This is followed by a conceptual evaluation phase and subsequent preparation of a basis of design report, which is followed by preparation of the design documents. The following sections present a suggested design sequence and description of the major design elements.

Conceptual Design

Define the concentration of contaminants in the raw water and develop minimum, average, and maximum concentrations. For surface water supplies, seasonal variations should be considered and data obtained. The concentration of all primary and secondary regulated contaminants should be obtained in order to ensure that all of the potential regulatory issues are addressed.

In addition to iron and manganese, analyses of the following constituents are also recommended, since they could impact the design and operation of the system:

- Hydrogen sulfide (H_2S)
- Silicon dioxide (SiO_2)
- Ammonia
- Radionuclides including radium, and uranium. Note that even if these values are below the MCL, some treatment media can remove these contaminants and the media may become radioactive.
- Contaminants that may be regulated by USEPA or are regulated by some states, e.g., perchlorate and methyl-*tert*-butyl-ether (M*t*BE)

In addition, for well water systems, a microscopic particulate analysis (MPA) should be conducted to assess if the groundwater is under the direct influence of surface water.

Define the treatment capacities of the system: minimum, average and maximum. Also establish any seasonal variations in capacity.

Define the treatment goals and regulatory requirements. For example:

- Is satisfying the secondary MCL acceptable, or are lower concentrations desirable?
- What are the regulatory requirements regarding loading rate? Is a waiver needed for the use of higher loading rates?

Define the site constraints including the following:

- Are there constraints on exposed vessels?
- Is the space site limited, requiring high-rate processes?

Define residuals-handling constraints including the following:

- Are sewers available, and do they appear to have sufficient capacity for iron and manganese liquid residuals?
- Is there sufficient space for sand drying beds?
- Are residuals required to be hauled off-site or recycled because of the lack of sewers?
- Is recycling of decanted residuals needed to minimize water consumption?

Based on these considerations, develop a set of treatment options for conceptual evaluation.

Perform a conceptual treatment evaluation that provides the following:

- Process floor plan showing overall equipment dimensions and orientation
- Site plan with concept contours and piping
- Estimate of electrical loads
- Capital, operating, and life-cycle costs based on process layouts
- Evaluation of noneconomic issues such as ease of operation and reliability

Discuss the evaluation with the owner, and provide a recommendation for final design.

Perform pilot testing if the process requires a loading rate higher than is permitted by regulations, if competing treatment technologies are being evaluated, or if there are site-specific water quality issues.

Detailed Design

Develop the basis of design report (BDR) that defines the design criteria for the treatment processes along with the design requirements for the support design disciplines (i.e., structural, architectural, electrical, etc.).

The BDR should include the following:

Design criteria

- Process and instrumentation diagram (P&ID). The equipment and piping portion of the P&ID should be included.
- A functional description of the control approach

- Process mechanical floor plan section completed to approximately 30 percent
- Site plan completed to approximately 30 percent
- A construction cost estimate

If permitted by regulatory agencies, consider submitting permit applications using the 30 percent drawings provided with the BDR. Discuss the BDR with the owner, and use it for development of the plans and specifications.

Design drawings

- Distribute the mechanical plan and section to other design disciplines and proceed with developing structural; architectural; plumbing; heating, ventilation and air conditioning (HVAC); and control design.
- Revise the process mechanical layout as needed.
- Develop technical specifications.
- Submit design drawings to the owner for review at the 60 percent, 90 percent, and 100 percent levels of design completion.

SUMMARY

The ultimate goal of an iron and manganese treatment system is to have a facility that does the following:

- Achieves the treatment objectives
- Is cost effective
- Is sustainable
- Allows for straightforward operation
- Is not complex and costly to maintain
- Enables comprehensive data collection along with the systems to analyze and interpret the data
- Is flexible enough to address future changes in regulations
- Will meet system needs with minimal upgrades for 20+ years

Satisfying these objectives will involve a project team consisting of design engineers, regulators, operators, and the owner. For the project to succeed especially in the long term, active participation of the owner and operators during the design is vital. These people have institutional knowledge regarding issues such as changes in source water quality, site conditions, and preferences regarding equipment operation and maintenance.

Throughout the design process, all the stakeholders should actively communicate by assessing and as necessary challenging the design decisions, equipment selection, and layout. Approaching the project design in a collaborative and open fashion will foster project success.

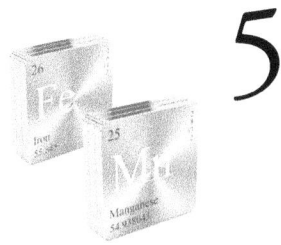

5

Oxidation

This chapter provides an overview of oxidation reactions, discusses the impact of temperature and pH on oxidation, describes the various forms of oxidation chemicals and systems, and concludes with a discussion of experiments that operators can use to measure the effectiveness of several forms of oxidation.

Oxidation is a chemical reaction in which electrons are removed from the atoms of a substance (e.g., iron) or a reductant being oxidized and transferred to atoms of an oxidant. Generally speaking, oxidation means loss of electrons from an atom, which in turn becomes oxidized and increases in positive valence.

Oxidation processes play an important role in drinking water treatment processes. Oxidation of metals such as ferrous iron and manganous manganese to insoluble forms permits their subsequent removal by physical/mechanical means. The rate at which oxidation reactions occur can be highly dependent on pH. Temperature can have some effect as well. The most common oxidants used in water treatment are chlorine, chlorine dioxide, potassium permanganate, and ozone.

Oxidants have some additional benefits in water treatment. They may improve the performance of some coagulants, control nuisance aquatic growths such as algae, and serve as primary disinfectants to meet primary disinfection contact time (CT) requirements.

The following sections discuss several oxidation reactions that may be used in the treatment of iron and manganese. A detailed understanding of the chemical reaction equations that follow is not necessary, but a basic understanding of the conditions required for these reactions to occur is necessary in order to design an effective oxidation process.

Optimum chemical dosages can critically affect any iron and manganese removal process. Overdosing will result in unnecessary costs, and underdosing will result in incomplete oxidation and insufficient removal of iron and manganese. For example, manganous manganese is the simplest form of soluble reduced manganese; its symbol is Mn^{2+}. When fully oxidized, it changes to manganic manganese, with the symbol Mn^{4+}. In order to reach that 4+ valence, each milligram per liter of manganese requires 1.3 mg/L of chlorine (in the form of hypochlorous acid, $HOCl$), or 1.92 mg/L of potassium permanganate ($KMnO_4$).

The actual chemical oxidant demand of a raw water often exceeds the theoretical demand for treatment of iron and manganese alone because other constituents, such as organic carbons and ammonia, contribute to oxidant demand.

The most commonly used oxidants in water treatment applications include:

- Air (atmospheric oxygen) as O_2 (aq) (meaning oxygen in a water solution, or aqueous oxygen)
- Chlorine (Cl_2) in the form of hypochlorous acid as $HOCl$ (available as a compressed gas, liquid sodium hypochlorite tablets, and via on-site generation)
- Potassium permanganate as $KMnO_4$ (dry)
- Sodium permanganate ($NaMnO_4$) (liquid)
- Chlorine dioxide as ClO_2
- Ozone as O_3 (aq)

Table 5-1 provides the chemical dosage needed to oxidize 1 mg/L of iron and manganese for each of the oxidants.

pH ADJUSTMENT AND REACTION TIME

Chapter 3 discusses the importance of pH to iron and manganese removal in the section titled "pH Value." Because pH is measured in numbers ranging from 1 to 15, operators can easily fall into the trap of dismissing a difference of 1 pH unit or less as insignificant, especially when they consider smaller differences, such as comparing pH 7.0 to pH 7.3.

Remember, however, that pH is measured on a logarithmic scale; every change of 1 pH unit represents a tenfold difference. For example, pH 8 indicates 10 times more alkalinity than pH 7, pH 9 indicates 100 times more alkalinity than pH 7, and so on. Viewed from this perspective, the importance of pH in a treatment process takes on a whole new meaning.

Table 5-1   Chemicals, reaction formula, and dosage needed to remove iron and manganese

Metal/ Oxidant	Reaction	Dosage
Iron		
O_2 (aq)	$2Fe^{2+} + \frac{1}{2}O_2 + 5H_2O \rightarrow 2Fe(OH)_3(s) + 4H^+$	0.14 mg: 1 mg Fe
O_3 (aq)	$2Fe^{2+} + O_3 + 5H_2O \rightarrow 2Fe(OH)_3(s) + O_2 + 4H^+$	0.43 mg: 1 mg Fe
HOCl	$2Fe^{2+} + HOCl + 5H_2O \rightarrow 2Fe(OH)_3(s) + Cl^- + 5H^+$	0.64 mg: 1 mg Fe
$KMnO_4$	$3Fe^{2+} + MnO_4 + 7H_2O \rightarrow 3Fe(OH)_3(s) + MnO_2 + 5H^+$	0.94 mg: 1 mg Fe
ClO_2	$5Fe^{2+} + ClO_2 + 13H_2O \rightarrow 5Fe(OH)_3(s) + Cl^- + 11H^+$	0.24 mg: 1 mg Fe
Manganese		
O_2 (aq)	$Mn^{2+} + \frac{1}{2}O_2 + H_2O \rightarrow MnO_2(s) + 2H^+$	0.29 mg: 1 mg Mn
O_3 (aq)	$Mn^{2+} + O_3 + H_2O \rightarrow MnO_2(s) + O_2 + 2H^+$	0.88 mg: 1 mg Mn
HOCl	$Mn^{2+} + HOCl + H_2O \rightarrow MnO_2(s) + Cl^- + 3H^+$	1.30 mg: 1 mg Mn
$KMnO_4$	$3Mn^{2+} + 2KMnO_4 + 2H_2O \rightarrow 5MnO_2(s) + 4H^+$	1.92 mg: 1 mg Mn
ClO_2	$Mn^{2+} + 2ClO_2 + 2H_2O \rightarrow MnO_2(s) + 2ClO_2^{2-} + 4H^+$	2.45 mg: 1 mg Mn

The rate of manganese oxidation induced by $KMnO_4$ is influenced by pH and temperature. Manganese oxidation at pH values between 5.5 and 9.0 generally occurs within 10 seconds at a water temperature of 25°C (77°F). At 2°C (36°F), oxidation could take 2 min or more, which is considered a long time in small plants whose designs do not incorporate detention time. For many years, the rate of manganese oxidation by $KMnO_4$ was considered especially dependent on pH, so soda ash was added to the raw water to increase its pH from, say, 7.0 to 7.5–7.7. Increasing raw water pH continues in plants where every second counts.

Aeration of a raw water will likely have no direct oxidizing effect on manganese in solution. By stripping the carbon dioxide (CO_2) out of the water, however, aeration raises the pH, and that change could affect how quickly iron and manganese can be oxidized.

This handbook cannot address all the potential ramifications of altering pH, but any operator should be aware of any natural changes in raw water pH. Plants should prepare to increase or decrease the pH of treated water to improve the overall quality for the end user and to minimize distribution system corrosion.

The oxidant addition will also consume alkalinity. The reduction in alkalinity could impact the corrosivity of the water and downstream treatment processes. Oxidants should be jar tested to determine if alkalinity adjustment is necessary.

Detention time and temperature are also important factors in determining the effectiveness of an oxidant. While groundwaters generally have stable temperatures, surface waters can have wide temperature variations. To compensate for changes in temperature, the operator can potentially change the dosage of the oxidant or add another chemical such as a coagulant.

For new plants or plants undergoing an upgrade, a preoxidation tank will provide additional operational flexibility and could potentially reduce the dosage of pH adjustment chemicals. An overview of jar testing procedures is provided at the end of this section. Jar testing will allow operators to optimize chemical dosages.

Tables 5-2 and 5-3 provide a summary of the theoretical oxidant dosage, optimal pH, and minimum reaction time needed for each of the oxidants.

Table 5-2 Summary of iron oxidation reactions

Oxidant	Oxidant needed to oxidize 1.0 mg Fe^{2+} (mg)	pH	Reaction time
Oxygen	0.14	>7.5	15 min
Chlorine	0.64	>8.0	15–30 min
Potassium permanganate	0.94	>5.5	<20 sec*
Ozone	0.43	–	Instantaneous
Chlorine dioxide	1.2	>5.5	10 sec

* Significantly longer when iron is complexed with natural organic matter.

Table 5-3 Summary of manganese oxidation reactions

Oxidant	Oxidant needed to oxidize 1.0 mg Mn^{2+} (mg)	pH	Reaction time
Oxygen	0.29	>9.0	>1 hr
Chlorine	1.30	>8.0	2–3 hr
Potassium permanganate	1.92	>5.5	<20 sec
Ozone	0.88		10–30 sec
Chlorine dioxide	2.5	>5.5	10 sec

AERATION

Aeration is often the first pretreatment measure used to prepare water for filtration. Water containing hydrogen sulfide is aerated to remove the gas from the water and release it to the atmosphere. Aeration also helps to oxidize iron. Oxygen, about 20 percent of air, oxidizes iron, though at varying rates.

Depending on water pH, temperature, and detention time, and in the absence of organic interference, oxidized iron often forms iron hydroxide, which agglomerates to a relatively large, heavy floc of particles that constitute filterable units.

The agglomerated iron is then filtered out in the upper portion of a filter bed made up of coal that is sized specifically for the water quality of the site, ranging in effective size from 0.7 to 1.2 mm (0.03 to 0.04 in.). Experience indicates that coal of this size is usually fine enough to filter out most iron hydroxides and create collisions of oxidized particles that enhance coagulation. At the same time, it is coarse enough to achieve a higher solids loading capacity than the sand or manganese greensand beneath it in most filter beds. If iron oxidized using aeration cannot be captured in a granular media filter, the operator can substitute a coagulant (alum, polymers, etc.) or an adsorption process.[1]

---

[1] Raymond Jones, Hungerford & Terry, 2015. Personal communication.

Forced-Air Cascading Tower

The most common method of aeration is the forced-air cascading tower. Water enters the tower from the top and cascades down over staggered wooden slats while a blower forces air up through the tower from the bottom. The water exits the tower into a detention tank from which it is pumped or allowed to overflow into the filters.

Cascading Steps

A variation of the cascading tower is cascading steps (Figure 5-1). In this configuration, water is simply allowed to flow down a channel that resembles a wide set of stairs open to the air. Sufficient turbulence results to dissolve enough oxygen into the water. Another variation is the water spray technique, in which water is sprayed into a detention tank, picking up oxygen from the air in the process.

Porous-Tube Pressure Aerator

Common in small treatment plants is the porous-tube type aerator Figure 5-2). Compressed air is forced into the center of a porous tube installed in a raw-water line. The air exits the tube in the form of millions of tiny bubbles that are picked up by the water as it flows. This method is effective,

Figure 5-1   Cascading aerator
Courtesy of Layne

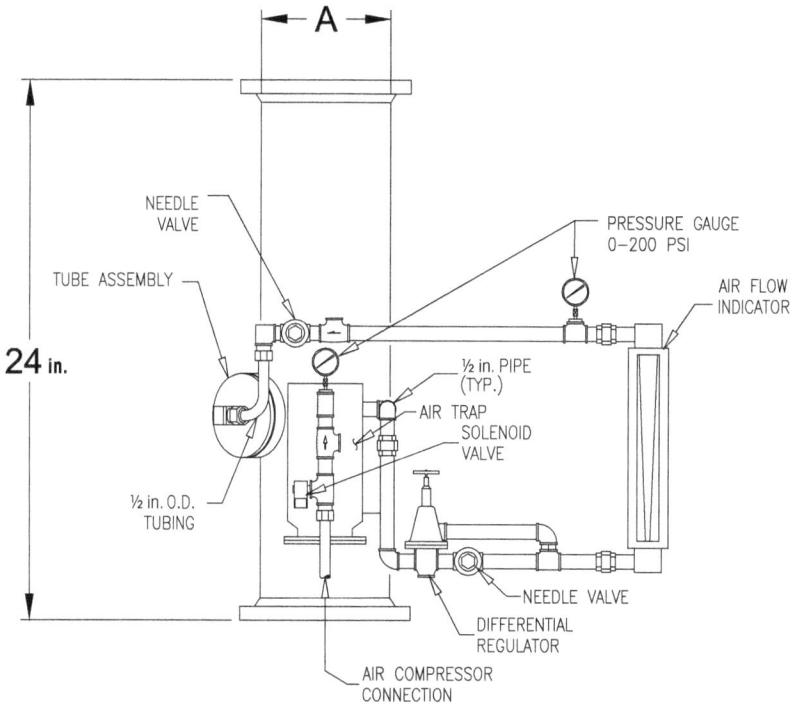

Figure 5-2   Porous-tube pressure aerator
Courtesy of Layne

but requires considerable maintenance since the needle valves controlling the air flow, compressor controls, and bleed lines require frequent servicing. Neglect of needed service often results in sporadic operation, which in turn results in inconsistent oxidation of iron. Failure to determine and follow an appropriate tube-cleaning schedule also results in inconsistent oxidation.

Venturi Device

Even smaller plants commonly use a series of Venturi devices (sometimes referred to as hydro-chargers). The Venturi principle takes advantage of suction created by water flowing at high velocity past a small orifice. A simple adjustment screw controls the amount of air drawn into the water flow. These devices are reasonably effective when correctly adjusted, but they require frequent cleaning to remove iron scale and rust buildup.

Aeration is generally suitable for the following conditions:

- Where only iron is present and there is greater than 15 min of detention time
- Where both iron and manganese are present at levels generally above 5 mg/L, which requires high dosages of oxidants such as chlorine or potassium permanganate. The operating cost of an aerator is generally lower than other chemical oxidants.
- Where raw-water organics are present at a level at which the addition of chlorine could result in elevated levels of disinfection by-products

CHLORINATION

Chlorine (Cl_2) is usually dosed in one of two ways: as a gas forced into water under pressure or as a solution pumped into the water line by a chemical pump.

Chlorine gas provides 100 percent available chlorine. Calcium hypochlorite is shipped in granular or pellet form and is mixed with water before dosing. The granules or pellets contain 65 percent available Cl_2. Sodium hypochlorite is shipped as a liquid and contains 12 percent available Cl_2. It is dosed either directly from the shipping barrel full strength or after dilution with water in a batch-mix tank.

Mixing chlorine with water forms hypochlorous acid, HOCl. When chlorine is mixed with hard water, sodium hydroxide (caustic) is also formed, which softens the water and produces precipitates. Many well waters exhibit high hardness and produce troublesome amounts of precipitates when softened. Mixing either of the hypochlorite products with water should take place in a batch tank, and time to settle and stabilize should be allowed before the diluted solution is drawn off into a second chemical container by a chemical pump. Without the precaution of the two-tank method, the precipitates can be drawn into the chemical pump and cause it to malfunction, or the precipitates can actually form inside the chemical pump and foul its operation.

Enough chlorine is usually dosed ahead of the filters to provide for oxidation of iron and a free chlorine residual of not less than 0.1 mg/L in the water leaving the filters. The dosage is usually 0.64 mg/L chlorine for each milligram per liter of iron in the raw water, plus the free chlorine residual, plus any other chlorine demands appropriate for the raw water. A

Cl_2 dosage substantially higher than the theoretical dosage indicates that the raw water has a hidden high demand for chlorine.

For example, water with a raw-iron content of 5 mg/L has an initial Cl_2 demand of 3.2 mg/L (5×0.64 mg/L, the theoretical iron demand), plus possible other minor demands and a free Cl_2 residual of 0.1–0.3 mg/L. The total theoretical dosage is about 3.5 mg/L (using a residual of 0.3 mg/L). If, however, a dosage of 5 mg/L is needed to oxidize the iron (Fe) and provide a free Cl_2 residual of 0.3 mg/L, what accounts for the additional demand?

In surface water, the additional demand often results from the presence of organic materials. In groundwater, the additional demand often reflects the presence of ammonia and hydrogen sulfide (rotten-egg-smelling gas). Where both of these compounds are found, organic carbons are also usually found. The chemical reaction between iron and chlorine is preferential to that between organic carbons and chlorine. Iron and chlorine typically (though not always) react to completion in under 1 min, while organic carbons and chlorine typically react to completion in 15 min (although the reaction sometimes takes hours). So if just the right dosage of chlorine needed to oxidize iron is used, none is left to react with organic carbons.

According to Knocke et al. (1990, p. 52), "Free chlorine was much more efficient in its ability to oxidize uncomplexed Fe(II). ... Although the reaction was not instantaneous ... efficient Fe(II) oxidation was typically observed within 10 to 15 s (seconds), even under pH 5.0 conditions. ... A decrease in solution temperature resulted in a noticeable decrease in Fe(II) oxidation rate. ... However, even under the lowest temperature conditions examined, effective Fe(II) oxidation by free chlorine was observed within 15 s (at pH 6.0) or 90 s (at pH 5.0)." This research was conducted using water temperatures of 2°C (36°F), 10°C (50°F), and 25°C (77°F).

While gas chlorine, bulk (12%) sodium hypochlorite, and calcium hypochlorite have been traditionally used for the application of chlorine, another chlorination option becoming more commonplace is the use of on-site generation of sodium hypochlorite (discussed in depth in AWWA M65, *On-Site Generation of Hypochlorite*). Gas chlorination is generally the least expensive chlorination method, but the storage of gas chlorine requires the use of enhanced safety measures, and these measures become more intensive when the gas storage requires the use of 1-ton containers.

Bulk sodium hypochlorite is much more expensive than gaseous chlorine (generally 3 to 4 times higher per pound), and although generally safer than gas chlorine, the high chemical concentration is subject to degradation and outgassing. Calcium hypochlorite is generally easier to feed than sodium hypochlorite but the cost is generally 2 to 3 times more than sodium hypochlorite.

An on-site generation system uses salt and electricity to generate a sodium hypochlorite solution with a strength of approximately 0.8 percent. These systems can produce sodium hypochlorite at a cost that is approximately twice the cost of gas chlorine. On-site systems are typically safer to operate than gas chlorine systems, as they do not require the storage of chlorine cylinders. On-site systems, however, typically require much more maintenance than gas chlorine systems.

Chlorine in any one of its forms is generally suitable for the following conditions:

- For most groundwater where the water is low in organic matter and the addition of chlorine will not cause disinfection by-product MCLs to exceed established limits
- For surface water plants that employ clarification and dual-media filtration, where the addition of chlorine downstream of clarification can be useful in oxidizing the manganese and continuously regenerating the manganese dioxide coating that forms on the sand particles
- Where breaking head (as required by aeration) is undesirable

PERMANGANATE

Permanganate is commonly used to oxidize both iron and manganese. It can also be used to regenerate manganese greensand or pyrolusite filter beds. Permanganate is available as a solid in the form of potassium permanganate ($KMnO_4$) and as a liquid as sodium permanganate ($MnNaO_4$). Potassium permanganate usually comes as purple crystals in 55-, 110-, and 330-lb (25-, 50-, and 150-kg) drums. The active ingredient is 95 percent to 99 percent strength, depending on the supplier.

Proper Storage for Potassium

Potassium permanganate reacts vigorously with organic materials, such as powdered or activated carbon, oil, and greases. Potassium permanganate should be stored away from other chemicals to avoid potentially violent reactions. Under some conditions, certain combinations can produce spontaneous explosions. In moist or humid conditions, potassium permanganate forms solid lumps, which will result in difficulties in creating the dilute solution necessary for feeding.

After removing sufficient chemical for a new batch, the plastic bag inside the drum should be secured to isolate the chemical from moist treatment plant air. If several drums are purchased at one time, the unopened ones should be stored in dry surroundings on a concrete floor. Potassium permanganate is always dissolved in solution before it is dosed. Typical potassium permanganate preparation and dosing systems for large plants are shown in Figure 5-3.

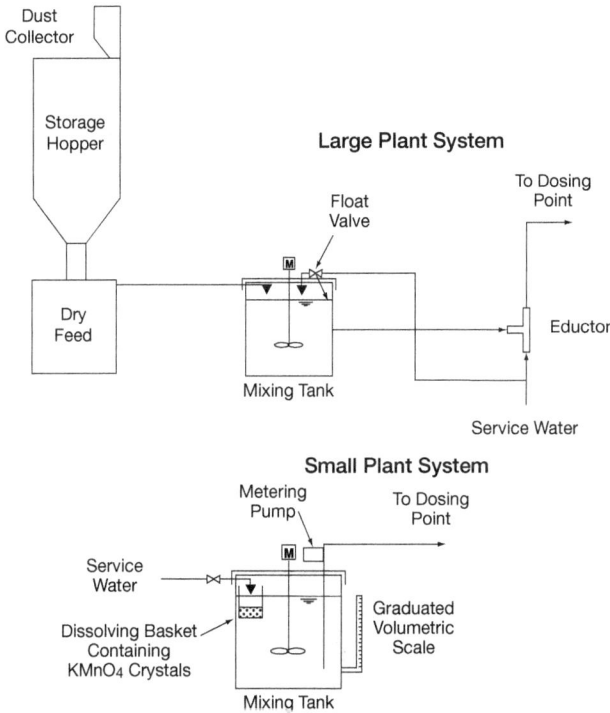

Figure 5-3 Potassium permanganate preparation and dosing systems
Courtesy of Reid Crowther & Partners Ltd.

Dosing

In a large plant, a volumetric feeder continuously adds a measured amount of crystals into a solution tank equipped with a mixer. An eductor draws the prepared solution from the tank and transfers it to the dosing location. Since the flow drawn from the mixing tank by the eductor is usually constant, dosing changes are made by adjusting the feed rate from the volumetric feeder.

This adjustment changes dosing by changing the solution strength. The same volume of solution is fed to the treatment process, but the active ingredient strength changes. For continuous dissolving systems like this, the solution strength should not exceed 1 percent.

In a small plant, batch solutions of known strength are mixed as required, usually twice each week or so. The solution is then fed to the raw water using a small chemical metering pump. Changes to the dosing rate (which is the same as the feed rate) are made by adjusting the metering pump. Solution strength is usually in the range of 1–3 percent by weight.

Solutions are prepared by simply pouring the $KMnO_4$ crystals directly into the batch tank as water is filling it or putting the crystals into a basket or porous sack that is then suspended in the batch tank. Many mixer designs are in use, the most common being a simple blade on the end of a shaft driven by an electric motor.

Water Temperature

The rate at which potassium permanganate goes into solution is powerfully influenced by water temperature. In room-temperature water (20°C/68°F), the crystals quickly dissolve. A typical small plant, however, uses water drawn from the clearwell at a temperature around 5°C (41°F). At this temperature, the slow dissolution rate requires the mixer to run for up to a half hour. The solubility/temperature curve for potassium permanganate in Figure 5-4 shows the dramatic effect of temperature on the degree of solubility.

The ideal water for batch mixing is good-quality treated water from the clearwell. Under no circumstances should raw water high in iron and manganese be used for batch mixing, as the iron and manganese will oxidize and precipitate out of solution. If precipitates build up, they will be drawn into the metering pump and piping, where they may cause blockage or poor pump performance. Residues that collect on the bottom of the batch tank should be removed from time to time.

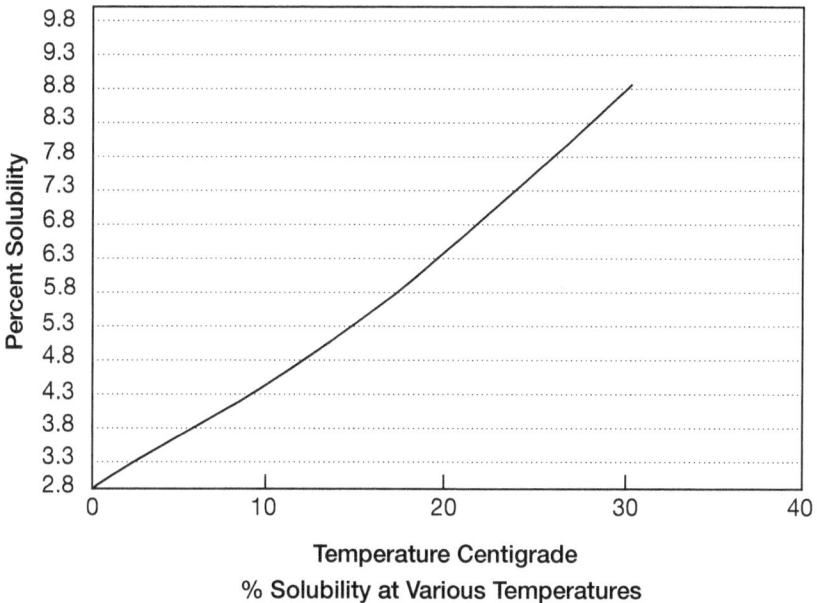

Figure 5-4   Increase in solubility of KMnO$_4$ with increased temperature

Two-Tank Method

If space and facilities permit, the best method is the two-tank method. The solution is mixed in one batch tank, allowed to settle, then transferred to a second tank without disturbing accumulated precipitates. The chemical metering pump draws solution for treatment from the second tank.

For a 1 percent solution, 1 kg (2.2 lb) of KMnO$_4$ is added to 100 L (26.4 gal) of water. For a 2 percent solution, the ratio is 2 kg to 100 L (4.4 lb to 26.4 gal). A common rule of thumb calls for 1 mg/L of KMnO$_4$ to oxidize each milligram per liter of iron and 2 mg/L KMnO$_4$ to oxidize each milligram per liter of manganese. (Table 5-1 provides the exact values.) But waters from different sources have different conditions, so be cautious when applying rules of thumb.

Liquid Sodium Permanganate

Another application method is the use of liquid sodium permanganate (MnNaO$_4$). Liquid sodium permanganate is available in both 20 percent and 40 percent concentrations. While MnNaO$_4$ is more expensive on a per-pound basis than potassium permanganate, it is simpler to feed and requires only storage tanks and metering pumps.

Permanganate is generally suitable for the following conditions:

- When the issue is primarily manganese
- When detention time is limited
- For initial conditioning of greensand and the newer types of manganese dioxide media
- For regeneration of conventional greensand
- For pretreatment of surface waters where prechlorination would produce elevated levels of disinfection by-products

CHLORINE DIOXIDE (ClO$_2$)

Oxidation rates for iron and manganese are extremely rapid—just a few seconds at room temperatures, and a bit slower at lower temperatures. In treatment with $KMnO_4$ and Cl_2, however, the presence of dissolved organic matter (i.e., organically complexed iron and manganese) tends to slow oxidation rates. Such a situation affects the use of chlorine dioxide (ClO$_2$) as an oxidant, since it differs from the other common oxidants in two ways.

First, chlorine dioxide reacts very rapidly with many organics, which makes the organic material an oxidant competitor to manganese. Extensive testing is required to determine the amounts and types of organic materials present in a raw water before dosage rates can be established. Second, a treatment plant using chlorine dioxide should have some means of tracking increases and decreases in organic levels, not an easy activity for a small plant.

Knocke et al. found that the interaction between chlorine dioxide and reduced manganese yields chlorite. Concerns regarding the potential health implications of residual chlorite concentrations might limit the usefulness of ClO$_2$ as a Mn(II) oxidant. The MCL for chlorite is 1 mg/L and often the chlorite MCL limits the dosage of chlorine dioxide to a maximum dosage of approximately 2 mg/L (Knocke et al. 1990).

Chlorate is also a by-product of chlorine dioxide, and the USEPA is considering regulating chlorate. Consideration should be given to the potential for chlorite and chlorate formation when using chlorine dioxide. Figure 5-5 shows a chlorine dioxide generator.

Figure 5-5   Chlorine dioxide generator
Courtesy of Chemours Water Technologies

Chlorine dioxide is always used in a gaseous form. The product is normally generated at the point of use.

A bulk chlorine dioxide solution with a concentration of 0.8 percent is now available from suppliers in 250-gal totes. Each tote contains approximately 15 lb of chlorine dioxide. For a 100-gpm system feeding 1 mg/L of chlorine dioxide, a single tote would last approximately 14 days.

Chlorine dioxide is generally suitable for the following conditions:

- Where raw-water organic concentrations prevent the use of chlorine due to the formation of disinfection by-products and there is insufficient detention time for the use of permanganate
- Where taste and odor problems also occur with iron and manganese
- For complexed forms of iron and manganese

OZONE

The effectiveness of ozone (O_3) on oxidation of iron and manganese is well documented in research conducted by Knocke et al. (1990) at Virginia

Polytechnic Institute and State University on behalf of AWWA and the Water Research Foundation (formerly the AWWA Research Foundation). They found that O_3 oxidizes iron almost instantly under typical pH and temperature conditions, but some types of organic complexing result in very poor rates of iron oxidation during O_3 treatment. Reaction times for manganese oxidation are somewhat slower but still in the range of 10 to 30 sec. Manganese oxidation is also affected by pH and temperature, and manganese oxidation rates are significantly inhibited by humic materials (carbon compounds). In relatively high doses, however, O_3 does a good job of oxidizing many organic carbon compounds. This method has one disadvantage: It does not leave a residual as a continuing disinfectant, as Cl_2 does.

Clearly, extensive knowledge of a raw water's chemical makeup is required in considering ozone treatment. Even with that knowledge, jar testing is recommended to assess initial dosages and reaction times. Upon completion of jar testing, pilot testing is recommended to optimize the design conditions and assess the impact on downstream unit processes.

Ozone is generally suitable for the following conditions:

- Where raw water organic concentrations prevent the use of chlorine due to the formation of disinfection by-products and there is insufficient detention time for the use of permanganate
- Where taste and odor problems also occur with iron and manganese
- Where the use of chlorine dioxide would require high doses such that the chlorite MCL would be exceeded
- Where raw water bromide levels are low such that the bromate MCL would not be exceeded
- For complex forms of iron and manganese

The cost of producing ozone has not decreased enough, however, to make this treatment method economically feasible for small plants. The extreme reactivity of ozone also complicates dosage control.

TESTING EQUIPMENT

Treatment plant operators conduct tests on (or analyze) iron and manganese in water for several reasons. When treatment processes are necessary to remove excess iron and manganese, concentrations of iron and manganese must be monitored for control of the treatment process. Source water and treatment effluent should be monitored to alert the operator to a pending change in the treatment requirement. Monitoring the process effluent is necessary to ensure the treatment process, including chemical feed systems, is functioning properly.

Even when treatment to remove excess iron and manganese is not needed, it is prudent to monitor iron and manganese in source water and in water entering the distribution system. For example, manganese is a common false-positive interference in chlorine tests.

When iron compounds such as ferric sulfate are used in the treatment process as a coagulant, source water iron as well as plant-effluent iron should be tested. The plant-effluent iron should never exceed the iron in the source water. Increasing iron during the treatment process would indicate improper control of the coagulation process. (The same is true where aluminum coagulants such as aluminum sulfate are used. Effluent aluminum should never exceed the source water aluminum concentration.)

While some large utilities may have the resources to prepare their own reagents, most utilities will find it is most cost effective to purchase commercial, prepared reagents. When regulatory reporting of total iron or total manganese is required, it is necessary to perform a digestion of the sample prior to analysis. However, most iron and manganese tests are only for purposes of process control, as outlined previously. Many commercial formulations are capable of achieving total iron or manganese without digestions, and thus it is often not necessary to perform digestions for process control testing.[1]

Digestion uses acid and heat to break organo-metallic bonds and free ions for analysis. Multiple types of digestion procedures can be used. Some involve the use of nitric acid and sodium hydroxide and others involve the use of sulfuric acid and hydrogen peroxide. If regulatory reporting is required, a certified laboratory should be used for the analysis.

Commercial products are available as reagent sets for use with laboratory colorimeters and spectrophotometers as well as in portable kits employing visual color comparators or hand-held colorimeters and portable spectrophotometers.

---

[1] Frank Kaiser, Hach Company, 2015. Personal communication.

TEST KITS

Test kits for visual comparison are quite common. However, the measurement results for a visual comparator are influenced by the background lighting (i.e., daylight versus an incandescent light versus fluorescent lighting) as well as the ability of the user to properly interpret the color. Color perception varies from individual to individual even in the same ambient light conditions, and many individuals suffer varying degrees of color blindness. Thus, when possible, a portable colorimeter or spectrophotometer is preferred to visual comparison for all measurements made by a water utility.

Colorimeters or spectrophotometers eliminate the subjective color judgment of the individual. Thus the measurements are much more accurate and consistent from test to test and among a group of operators. Many of these devices are precalibrated for use with the manufacturer's reagent sets. It is important to closely follow the manufacturer's test instructions for measuring and mixing of sample and reagents, using the proper sample cells, and disposing of the sample at completion of the test. Many test procedures require a timed reaction period, and most protocol will contain information on possible interferences and procedures for dealing with interferences. Finally, for best results always use a test kit as designed by the manufacturer. Do not mix reagents from one manufacturer with a precalibrated instrument from another manufacturer.

METHODS OF TESTING FOR SOLUBLE METALS

Testing for Iron

Ferrous iron (Fe^{2+}) can be measured based on color changes developed by a number of reagents. One such reagent is 1,10-phenantroline (the indicator). The commercial formulation may also contain other reagents to establish proper sample conditions such as the pH. When Fe^{2+} comes into contact with phenantroline, a reddish-orange coloration develops with an intensity directly proportional to iron concentration, assuming an excess of phenantroline remains. Intensity of the color is measured with a colorimeter or spectrophotometer at a wavelength of 510 nanometers (nm). Handheld color comparators are also available for visual measurement.

Total iron may also be determined using 1,10-phenantroline (ferric iron, Fe^{3+}, plus ferrous iron, Fe^{2+}). First a reducing agent is added to change the ferric iron to ferrous iron (Fe^{3+} to Fe^{2+}) before measurement. Reagents for total iron will contain the indicator, reducing agent, and a buffer for pH. This reaction takes about 15 min. A titration method based on 1,10 phenanthroline is available for total iron for very high concentrations up to 10,000 mg/L.

Other common methods for determining total iron include ferrozine and 2,4,6-tripyridyl-s-triazine (TPTZ). As with measurement of total iron with 1,10 phenanthroline, commercial formulations of ferrozine and TPTZ iron reagents will contain reducing agents and appropriate buffering reagents. A special formulation of TPTZ is available for measurement of iron in cooling towers where molybdate tracers are used.

The ferrozine reagent is very sensitive and is used for parts per billion analysis of iron. One must be careful when using ferrozine not to contaminate the reagent. Many times operators will use metallic scissors or clippers to open reagent packages, which can cause a false-positive error in the test of ferrozine from the iron left by the scissors or clippers on the package.

When only ferric iron is desired, one should perform tests for both total iron and ferrous iron. The difference is ferric iron (total iron – ferrous iron = ferric iron).

Several substances may interfere with the determination of iron concentration. See the kit and reagent manufacturer's instructions for possible interferences and the proper procedure for dealing with the interferences.

Because of the variety of platforms (i.e., color comparator, colorimeter, spectrophotometer) and variety of reagents available, make certain the test system (kit) selected is appropriate for the sample and the expected range of measurement. Where possible, it is best to select a test range for which the sample to be measured will fall near the middle of the range of the kit.

Testing for Manganese

The presence of manganese in water typically is determined using one of two methods: the periodate method for high range or the PAN [(1-(2-pyridylazo)-2-napthol)] method for low range. As with measurement of iron reagent, sets and portable kits are available using visual color comparators, colorimeters, or spectrophotometers for both methods.

The periodate method is very simple. Manganese present in the sample is oxidized to Mn^{7+} by addition of periodate (IO_4). The oxidized manganese forms a color without the need for an indicator. Thus the intensity of the color can be measured directly at a wavelength of 525 nm.

In addition to the reducing agent, the reagent system may include a buffer and additional reagents for overcoming interferences in the test, which may include turbidity and organic matter. The periodate method is suitable for use from about 0.1 mg/L to 20 mg/L of manganese.

The PAN method is preferred for low-range measurements, about 0.006–0.700 mg/L. The PAN test uses three reagents: alkaline cyanide, ascorbic acid powder pillow, and PAN indicator solution. The color developed in the test is measured at a wavelength of 560 nm.

When using the PAN method kit, allow 10 min for full color development if iron in the sample exceeds 5 mg/L. If sample hardness as calcium carbonate ($CaCO_3$) exceeds 300 mg/L, add Rochelle salt solution as recommended by the kit manufacturer.

As with the iron test kits and reagents, make certain to follow the manufacturer's test procedure and consult the procedure for possible interferences, dealing with the interferences, and proper sampling and disposal. Always use good safety practices in handling of reagents, use of laboratory equipment, and disposal of waste. Consult the manufacturer's material safety data sheet (MSDS) for all reagents.

TESTING OXIDATION EFFICIENCY

As discussed, removal of iron and manganese from drinking water often entails precipitation by oxidation, then separation of the solids by granular media filtration. If water quality objectives for iron and manganese removal are not met, the level of oxidation is the first variable to check. If a high level of oxidation is taking place, then the precipitate must be passing through the filter, requiring further investigation.

Iron may precipitate as iron hydroxide and manganese as manganic dioxide. Both may appear in colloidal form. (A colloid is a dispersion of particles larger than those in a true solution and smaller than those in a true suspension.) The concentration of oxidized metals can be determined by filtering a sample through a membrane of known pore size before performing one of the iron or manganese tests described earlier.

An apparatus for membrane filtration analysis (sometimes called *millepore testing*) is shown in Figure 5-6. A membrane of known pore size is placed in the holder, and the sample is poured into the container above it. A hand vacuum pump removes enough air from the filtration flask to draw the sample water through the membrane, typically at less than 35 kPa (5 psi). Membrane filter disks (most of them round) come in pore sizes as small as 0.22 μm. A common size is 0.45 μm.

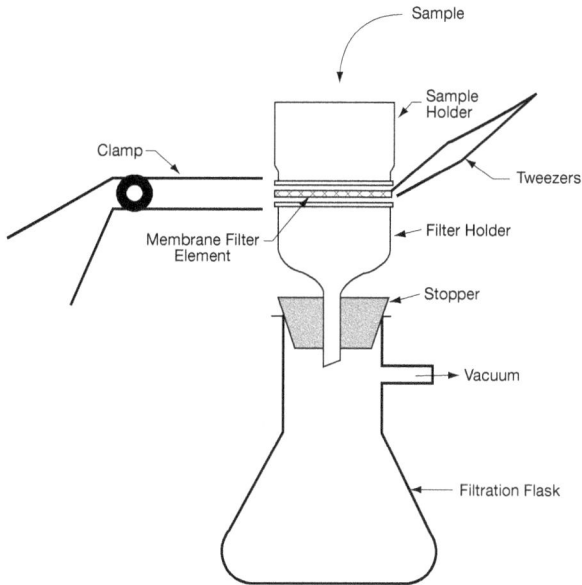

Figure 5-6   Dissolved metal analysis by membrane filtration
Courtesy of Reid Crowther & Partners Ltd

If a sample drawn through a 0.45-μm filter still has a high percentage of the raw water levels of iron and manganese, a granular filter will not remove them. If a similar sample results after dosing with an oxidant [O_2 (aq), Cl_2, or $KMnO_4$], the question is whether the particulates are very fine or colloidal in size, or if the iron and manganese are not oxidized. Sometimes both O_2 (aq) and Cl_2 can take from several minutes to hours to complete iron and manganese oxidation. Sometimes both Cl_2 and $KMnO_4$ can oxidize iron and manganese quickly, but the resulting precipitate may remain very fine or colloidal in size.

For further analysis, take a sample of oxidized raw water from the plant and add measured amounts of a coagulant, say alum or a cationic polymer, to a series of jars. Stir gently for several minutes (say 15 min), allow the samples to settle for several minutes (say 15 min more), then filter through a 0.45-μm membrane again. If the filtrate then tests very low in iron and manganese, the particles in the first sample were very fine or colloidal in size.

This result indicates that a coagulant (selected on the basis of further pilot testing) should be a part of the pretreatment process. In some instances, oxidized iron and manganese can even be subcolloidal in size and extremely difficult to flocculate, even using coagulants.

If the addition of a coagulant makes little or no difference, the iron and manganese likely remain in the sample because they were not oxidized. In this case, conduct tests on each of the three potential oxidants (i.e., O_2 [aq], Cl_2, and $KMnO_4$), allowing a reasonable amount of time for the oxidation reaction to go to completion. These tests should determine how long complete oxidation takes. The challenge then is to determine if that amount of time can be made available within the treatment sequence.

Most small plants can allow only very short detention times. Some plants have room to install tank capacity that can lengthen detention time by several minutes or up to a half hour. If detention time tests indicate a need for hours of detention time, some other treatment process should be considered.

Membrane filter tests using a number of different pore sizes can help to establish a particulate size distribution. For example, if the finest test membrane is 0.45 μm and the coarsest is 10.0 μm with two or three sizes in between, testing a sample of oxidized water may show a percentage of particulate removed by each pore size. These results are compared to the smallest particulate size that can be removed in the plant's filters.

TESTING FOR CHEMICAL OXIDANT DEMAND

Measuring chemical oxidant demand provides important information for correctly dosing chemicals for proper treatment. The demand test generally requires addition of the oxidant [usually O_2 (aq), Cl_2, or $KMnO_4$] in known quantities, a reaction time similar to what occurs in the treatment plant, and determination of the oxidant residual after completion of the reaction time. The oxidant demand is the difference between the dose added and the residual.

Oxygen (aq) Demand

Because an oxygen residual in water does not have an important effect on health, little effort is usually taken to determine just how much is required. A dissolved oxygen saturation level of about 60 percent is typically sufficient for iron oxidation. However, an operator may want to know just how much DO results from any aeration process. Too much could cause problems such as oxygen accumulation (bubbles) in an underdrain, which may result in media disruption during backwash.

Under certain circumstances, air could be forced out of solution in the filter media and reduce filter run lengths. Too much oxygen might also stimulate unwanted microbiological activity. In particular, an occasional measurement of DO is wise during compliance inspections, with adjustments to aerators made as necessary.

Chlorine Demand Example

For this example, four 1,000-mL samples of raw water are placed in four 1-liter beakers. The first receives 1 mL of calcium hypochlorite solution containing 1 g/L of Cl_2. The second beaker receives 2 mL of the same solution, the third 3 mL, and the fourth 4 mL. The beakers are then kept in the dark to prevent photochemical reactions.

If the plant's hydraulic detention time between chemical dosing and filtration is about 10 min, the samples are tested for total chlorine residual after that time. No chlorine residual is detected in the first beaker. The chlorine residual in the second beaker is 0.05 mg/L. The residual in the third beaker is 1.1 mg/L. The fourth sample is not analyzed.

To find the raw water's chlorine demand, first determine how much chlorine was added to each beaker. Because the dosage contained 1 g/L of Cl_2, the chemical equals 1,000 mg/L, or 1,000 mg/1,000 mL. The first beaker of sample received 1 mL of solution in 1,000 mL (1 liter). Therefore, the dosage of chlorine for the first beaker is as follows:

$$\text{amount of } Cl_2 \text{ added} = \text{amount of solution added}$$
$$\times \text{ solution concentration}$$
$$= 1 \text{ mL of solution} \times 1,000 \text{ mg } Cl_2 \text{ per } 1,000 \text{ mg of solution}$$
$$= 1 \text{ mg } Cl_2$$
$$\text{chlorine dosage} = \text{amount of } Cl_2 \text{ added/sample volume}$$
$$= 1 \text{ mg } Cl_2/1,000 \text{ mL of sample} = 1 \text{ mg/L}$$

The dosage for the other beakers can be calculated similarly as 2, 3, and 4 mg/L, respectively. The zero residual concentration of chlorine in the first beaker indicates that the chlorine demand in the raw water was at least 1 mg/L, but probably greater. The demand for the second beaker can be calculated as follows:

$$\text{chlorine demand} = \text{chlorine dosage} - \text{chlorine residual}$$
$$= 2 \text{ mg/L} - 0.05 \text{ mg/L} = 1.95 \text{ mg/L}$$

The chlorine demand for the third beaker can be calculated similarly as 1.9 mg/L. The similarity of the chlorine demand measured in the second and third beakers confirms that the raw water has a Cl_2 demand of 1.9 mg/L. The average of the two could have been taken, but they are obviously so close that this step is not needed. The sample with the greatest residual concentration (i.e., 3 mg/L) is likely the one with the least error. Because the second and third beakers indicated the same chlorine demand, the fourth sample need not be analyzed. The chlorine demand in the raw water is established as 1.9 mg/L.

Reagent Solution Preparation

To prepare a Cl_2 reagent solution with a concentration of 1 g/L, 1.55 g of calcium hypochlorite is dissolved in 1,000 mL of distilled water contained in a clean, dark bottle. This solution should be prepared immediately before use in the test, and any remainder should be discarded after all tests have been completed for the day. Calcium hypochlorite is unstable and cannot be stored for a long time. (Sodium hypochlorite could be used, but it is an even more unstable compound.)

Potassium Permanganate Demand Example

To determine potassium permanganate demand, follow the process of the chlorine demand test in the previous example. However, instead of testing for residual $KMnO_4$, coloration is observed after the reaction time. Oxidized iron and manganese form a reddish-orange to brown precipitate. Residual $KMnO_4$ gives the solution a pink coloration. Therefore, the sample demand falls between the dosages corresponding to the two consecutive beaker samples in which one has no pink coloration and the next one has a pink coloration. To fine-tune the demand determination, repeat the test using dosage levels between the two used in the identified beakers.

For example, if the first test found a $KMnO_4$ demand between 1 and 2 mg/L, the next test could use dosages of 1.2, 1.4, 1.6, and 1.8 mg/L. If the one dosed with 1.2 mg/L $KMnO_4$ shows no pink, but the one dosed with 1.4 mg/L does show pink, the demand could be reported as 1.3 mg/L. On the other hand, if all beakers showed a pink coloration, the demand would be 1.1 mg/L. (Remember that the beaker with 1 mg/L was not pink in the first test, so the demand must be greater than that amount.)

To prepare a 1-g/L reagent solution of $KMnO_4$, dissolve 1.05 g of $KMnO_4$ in 1,000 mL of distilled water in a clean, dark bottle. This bottle should have a plastic cap, a label indicating its contents, and the date of preparation. This solution can be used for about a month if kept in a cool, dark place.

SUMMARY

Oxidation of iron and manganese is the first step in the removal process. The choice of the oxidant will be based on an evaluation of issues such cost, site considerations, source water quality, downstream clarification and filtration processes, and system complexity. Testing for iron and manganese as well as for oxidant demand can be performed on site for initial evaluations as well as for process optimization. The on-site testing should be supplemented by analyses by a certified laboratory.

6

Clarification

For source water with high iron and manganese concentrations (above 5–10 mg/L), filtration technologies generally result in a significant amount of backwash water and thus operate at low water efficiency due to short filter run times. In such cases, clarification can help remove the bulk of the iron and manganese and improve the efficiency of the filtration process. Traditionally, clarification has been limited to sedimentation, and in some instances solids-contact clarification. Over the past couple of decades, however, advancements have been made to plate and tube settlers, which has made these high-rate systems viable clarification treatment technologies. In some instances, high-rated technologies such as ballasted flocculation can also be considered. This chapter presents practical design information regarding clarification technologies.

BENEFITS OF CLARIFICATION

Consider a groundwater supply with 4 mg/L of iron and 1 mg/L of manganese with a capacity of 700 gpm. The system operates 24 hr/day, 7 days per week. A traditional greensand-type pressure filtration system (discussed in chapter 7) would have a typical filter run time of 6.5 hr and generate approximately 21 mil gal of spent filter backwash annually.

If a clarification system is provided upstream of the greensand pressure filtration system, the filter run time would increase to 20 hr and the spent filter backwash volume would be reduced to approximately 4.7 mil gal. While the addition of a clarification system will increase capital costs, backwash water production will be significantly reduced and the savings in residuals handling costs will more than offset the additional construction costs associated with a clarifier.

Generally, a clarification system is not cost effective when the combined iron and manganese total less than 5 mg/L because the cost of clarification is not offset by the savings in residuals production. Because site conditions and residuals handling and disposal costs have a significant impact on whether a clarification system is beneficial, a cost-benefit analysis should be performed for each project.

CLARIFICATION FUNDAMENTALS

Once a decision has been made to use clarification, a number of clarification options need to be evaluated. All clarifiers operate according to fundamental fluid mechanics principles. Figure 6-1 presents a diagram of a particle in water. The particle is acted on by drag and buoyant forces. The drag force is affected by the area of the particle and the density of the water. The buoyant force is affected by volume of the particle, density of the particle, and density of the water.

For clarification processes, the settling velocity becomes the primary parameter for design and operations. The settling velocity is usually represented in units of gallons per minute per square foot (gpm/ft^2) and is often referred to as a basin loading rate. For oxidized iron and manganese particles, a settling velocity of approximately 0.4 gpm/ft^2 is often representative of typical conditions. The units of gpm/ft^2 are actually a velocity and can be represented in feet per minute (ft/min). Jar tests are often performed to measure site-specific settling velocities, and the results of jar tests are typically reported in centimeters per minute (cm/min). The conversions from gpm/ft^2 to ft/min and cm/min are shown below.

$$0.4 \text{ gpm/ft}^2 \times 1 \text{ ft}^3/7.48 \text{ gal} = 0.053 \text{ ft/min}$$

$$0.053 \text{ ft/min} \times 12 \text{ in./ft} \times 2.54 \text{ cm/in.} = 1.62 \text{ cm/min}$$

For an existing plant with a conventional clarifier, the basin surface loading rate can be computed by dividing the basin surface by the operating flow rate. A jar test can be performed to measure the settling rate. If the settling rate of the jar test is greater than the basin loading rate, the clarifier should be operating properly. If the settling rate is lower than the basin loading rate, modifications to the pretreatment process should be considered, such as the addition of a coagulant to increase the settling rate. For a new plant, jar tests should be conducted as part of the conceptual design to determine the design loading for the clarification system.

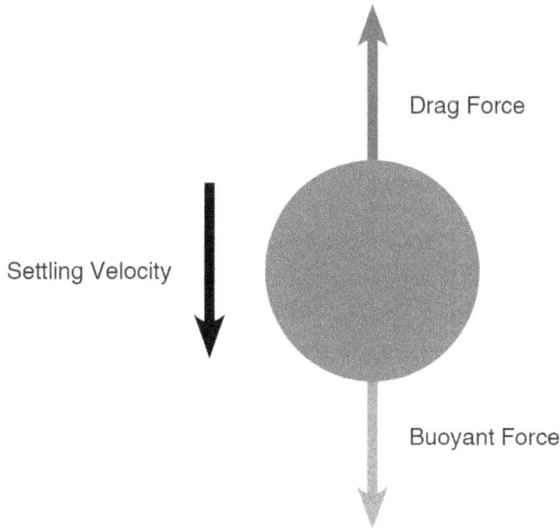

Figure 6-1   Fluid mechanics of particle settling

Temperature effects are generally more pronounced in surface waters, and testing with the minimum water temperatures is recommended. To overcome the potential adverse impacts of water temperature, the density of the particle can be modified. Coagulants and coagulant-aid chemicals are available that can result in denser particles that settle faster. Dissolved air flotation is used (discussed later in this chapter) when a lighter, more buoyant floc is desired and coagulants are available to accomplish this purpose. Another option to increase the settling velocity is the addition of a ballast that can increase the settling velocity by a factor of 10. This process, known as *ballasted flocculation,* is also discussed later in this chapter.

Types of Clarification

As discussed in chapter 4, oxidation is required to convert the iron and manganese from their soluble form to their particulate form. For most groundwaters, this conversion can be accomplished with the addition of a chemical oxidant and associated detention time. Surface waters generally have iron and manganese that are complexed with organic material. Because of the presence of organics, a coagulant may be needed and flocculation is often necessary to form a larger, heavier particle that can be settled. The types of clarification systems used for removal of iron and manganese include the following:

- Conventional sedimentation basins
- Plate and tube settlers
- Solids-contact clarifiers
- Dissolved air flotation
- Ballasted flocculation

The design of each of these systems is based on the requirements of the state or provincial regulatory authorities along with the site-specific water quality conditions. Despite the numerous discrete regulatory authorities, a general consensus exists regarding the design requirements. This section provides an overview of the features of each of these clarification systems as they relate to iron and manganese treatment.

Conventional Sedimentation Basins

Conventional sedimentation basins are typically rectangular with a length-to-width ratio of approximately 4 to 1 and a depth of 10 to 15 ft. The surface loading rate of a conventional sedimentation basin is usually 0.4 to 0.5 gpm/ft², with a detention time of approximately 4 hr. A sedimentation basin has several components, each of which needs to be sized. These components and design guidance include the following:

Inlet devices. Inlet devices are used to distribute the water uniformly in the basin. A baffle is often used to dissipate inlet velocities.

Outlet devices. Outlet weirs or submerged orifices are used to collect clarified water. The rate of flow is typically limited to 20,000 gpd/ft of the outlet launder.

Velocity. The velocity through the settling basins should be limited to a maximum 0.5 ft/min.

Flocculation. If the source is a surface water, then flocculation will be needed, with a minimum of two stages and a minimum total detention of 30 min.

Sludge collection system. The clarifiers can be manually cleaned or a mechanical collection system can be installed. If the basin is to be manually cleaned, basin bottoms should slope toward the drain not less than 1 ft in 12 ft.

Grit, corrosion, and clogging can subject mechanical collection systems to harsh operating conditions and wear, often resulting in the need for frequent maintenance. These mechanical systems have become more robust in recent years through the use of more durable components. Several types of sludge collection systems can be installed in rectangular sedimentation basins, including the following:

- Chain and flight type
- Perforated pipe type
- Scraper type

These systems can often be retrofitted into an existing basin.

Plate and Tube Settlers

Plate and tube settlers are used to reduce the amount of land required (as measured by plan surface area) for clarification. Plate and tube settlers generally require less than one-quarter the surface area of a conventional sedimentation basin. The construction cost is typically 70 percent of a conventional sedimentation basin.

With plate and tube settlers, the settling characteristics of the particle remain unchanged. The reduction in surface area is achieved by installation of plates or tubes in a basin, which effectively creates multiple clarifiers. Table 6-1 presents the concept behind plate and tube technology.

The differences between plate and tube settlers is primarily in their spacing, configuration, and construction materials. Regardless of which technology is selected, the fundamental design and operating parameter is the settling rate, which is generally limited to 0.4 to 0.5 gpm/ft^2. Plate and tube settlers can be built as new systems or can be retrofit into existing clarifiers to increase capacity.

Plate settlers. Figure 6-2 presents a plan and section of a plate settler. Plate settlers for the treatment of iron and manganese typically use a plate with a dimension of 4.5 ft × 10 ft and are inclined at an angle of 55 degrees and are spaced approximately 2 in. apart. The construction material is generally stainless steel, although other materials such as fiberglass reinforced plastic (FRP) are available. The plates are removable and an individual plate typically weighs about 57 lb. As long as they are properly maintained, the plates will last more than 30 years.

Table 6-1 Concept behind plate and tube technology

	24-in. Tube	36-in. Tube	48-in. Tube
Projected Area of Tube (per ft of basin width)	2 ft (cos 60°) × 1 ft width = 1.0 ft²	3 ft (cos 60°) × 1 ft width = 1.5 ft²	4 ft (cos 60°) × 1 ft width = 2.0 ft²
Number of Tubes (per ft of basin width)	(12) in. / (2 in. /sin 60°) = 5.19 tubes/ft of basin width		
Therefore there are:	5.19 ft² of projected tube area per 1 ft² of basin area for a 2-in. tube	7.79 ft² of projected tube area per 1 ft² of basin area for a 3-in. tube	10.38 ft² of projected tube area per 1 ft² of basin area for a 4-in. tube
Projected Effective Tube Area (ETA)	3.24 ft² for a 2-in. tube [(15/24) × 5.19]	5.84 ft² for a 3-in. tube [(27/36) × 7.79]	8.43 ft² for a 4-in. tube [39/48) × 10.38]
Calculation of Settling Velocity	If basin is operated at 1.5 gpm/ft² of basin area (BA), the settling velocity will be 1.5 gpm/ft² of BA/3.24 ft² of tube area = 0.46 gpm/ft² of ETA (1.5/3.24)	If basin is operated at 2.5 gpm/ft² of BA, the settling velocity will be 2.5 gpm/ft² of BA/5.84 ft² of tube area = 0.43 gpm/ft² of ETA (2.5/5.84)	If basin is operated at 3.5 gpm/ft² of BA, the settling velocity will be 3.5 gpm/ft² of BA/8.43 ft² of tube area = 0.41 gpm/ft² of ETA (3.5/8.43)
Manufacturer's Recommended Loading	1.5 gpm/ft²	2.5 gpm/ft²	3.5 gpm/ft²
Regulatory Guidance	Typical particle settling velocity = 0.5 gpm/ft². Regulatory standard for tubes is generally 2 gpm/ft². Manufacturer's recommended loading rate for tubes (gpm/ft² of basin) results in a particle settling velocity of less than 0.5 gpm/ft². Coordinate loading rate with appropriate regulatory authority.		
Adaptability	Similar calculations can be performed for plate settlers using plate dimensions/spacing.		

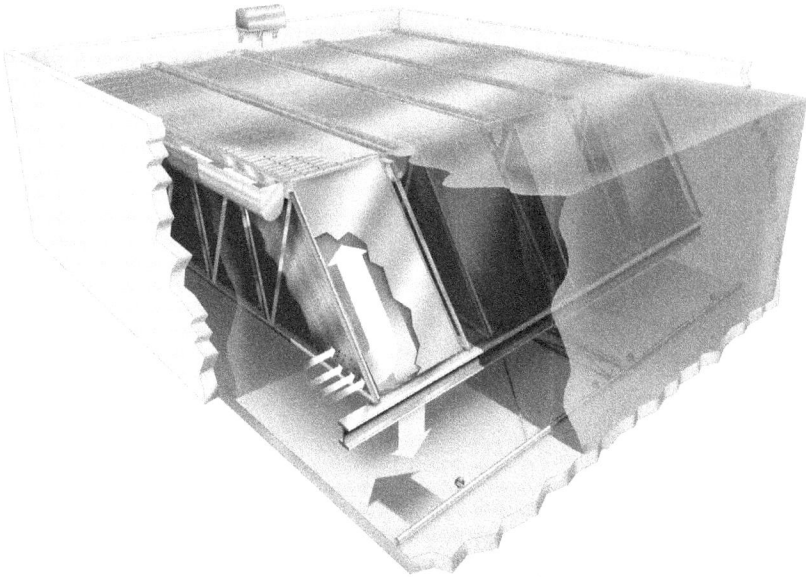

Figure 6-2   Plan and section of a plate settler
Courtesy of Meurer Research

Plate settlers are generally designed based on the following:

Application rates. Base application rates on a maximum plate loading rate of 0.5 gpm/ft² (1.2 m/hr), based on 80 percent of the projected horizontal plate area.

Inlet and outlets. Design these to maintain velocities suitable for settling in the basin and to minimize short-circuiting. It is critical that the plate units be designed with adjustable weirs to minimize misdistribution across the units. When selecting a plate system, consider contacting and visiting an installation with similar conditions as those being considered.

Drainage piping. Size drainage piping from the settler units to facilitate a quick flush of the settler units and to prevent flooding other portions of the plant. An overflow in the form of a weir or pipe should be provided.

Structural support system. Space is needed for a structural support system to carry the weight of the modules when the basin is drained and any additional maintenance-related weight. Although the plates are at an angle of 55 degrees, floc particles will adhere to the plates and the plates will require periodic cleaning. Periodic cleaning is generally accomplished by lowering the water level and hosing off the plates. High-pressure water can be used only if the plate material can withstand the pressure. Stainless steel plates can withstand a water pressure up to 50 psi and can also withstand the live load of an operator walking on the units.

Tube settlers. Most of the design criteria described for plate settlers are applicable to tube settlers. However, tube settlers are generally available in polyvinyl chloride (PVC) and polyethylene acrylonitrile butadiene styrene (ABS) and are generally not available in stainless steel. Tube settlers are generally less expensive than stainless steel plates. Due to the construction materials, tube settlers are generally replaced approximately every 10 years.

The regulatory design criteria generally limit the application rate to 2 gpm/ft^2 of basin area. However, as discussed earlier, the tubes do not change the settling characteristics of the particle. Tube settlers are commercially available in 2-, 3-, and 4-ft lengths. The longer lengths result in more available settling area for a given basin size.

When sizing tube settlers, establishing a settling velocity using jar tests is recommended. In the absence of jar testing, a maximum settling of 0.5 gpm/ft^2 should be used. The required projected horizontal tube area should then be calculated, and the length and quantity of tubes should be selected to provide the horizontal tube area.

Solids-Contact Clarifiers

A solids-contact clarifier combines flocculation and clarification into a single treatment process. The solids-contact unit contains a mixing zone, a reaction-flocculation zone, a sludge-blanket zone, and an outlet zone. Figure 6-3 presents a sectional view of a solids-contact clarifier.

The surface loading rate is the primary design parameter. Some regulatory agencies limit the loading rate to 1.0 gpm/ft^2 at the sludge separation line, whereas other jurisdictions limit the loading rate to 1.0 gpm/ft^2 at a level 5 ft below the level of the discharge weir/orifices. Coordination is important with the manufacturer for the location of the sludge separation line and with the regulatory authority for the approved design loading rate.

Dissolved Air Flotation

Dissolved air flotation (DAF) is not in itself a process for removing iron and manganese. However, like sedimentation, it does remove the precipitates formed following the oxidation of the dissolved iron and manganese by agents such as aeration, Cl_2, $KMnO_4$, and O_3.

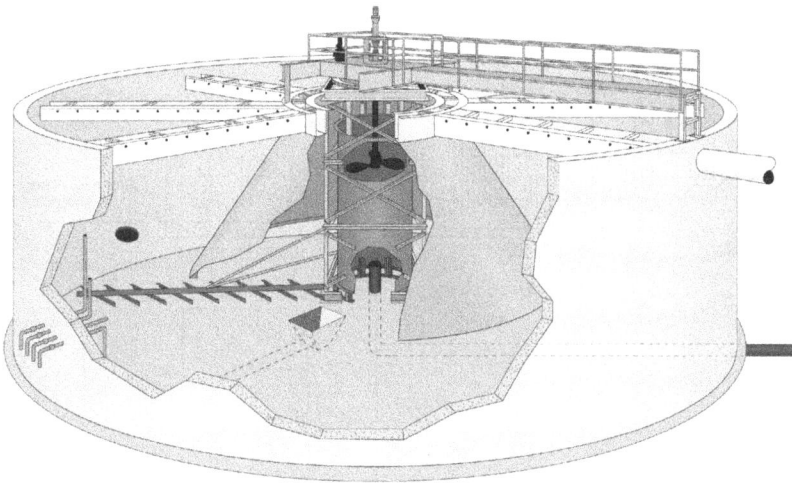

Figure 6-3   Solids-contact clarifier
Courtesy of WesTech Engineering Inc.

A typical DAF process is shown in Figure 6-4. The iron and manganese are first oxidized, and based on the type of oxidant, sufficient time is allowed for completion of the oxidation reaction. Next, a coagulant such as alum is added to the water, causing the iron and manganese precipitates produced by oxidation to become suspended in solution, where they clump together as floc. Gentle stirring promotes flocculation, and the floc grow over time to sizes that allow easy removal.

As the flocculated water enters the DAF cell, microscopic air bubbles are injected into the flow, where they attach to the particles of floc. The floc then rides to the surface on the air bubbles, where it forms a thick scum. At regular intervals, the scum is removed using rotating paddles. Clear water then leaves from the lower outlet end of the DAF cell.

Process efficiency demands extremely small air bubbles. To achieve this condition, clarified water from the DAF cell is pumped into a saturator, a device for mixing air and water under pressure of approximately 55–70 psi (roughly 380–480 kPa). The recycle rate is generally 6 percent to 10 percent of the clarified water. A flow of pressurized, air-saturated water flows from the bottom of the saturator to injection nozzles at the DAF cell inlet. As the flow is forced through the nozzles, the sudden pressure change causes the air to come out of solution in the form of very fine bubbles, similar to what occurs when the top is removed from a bottle of soda or beer. Shaking the bottle first creates massive bubble release and surface froth.

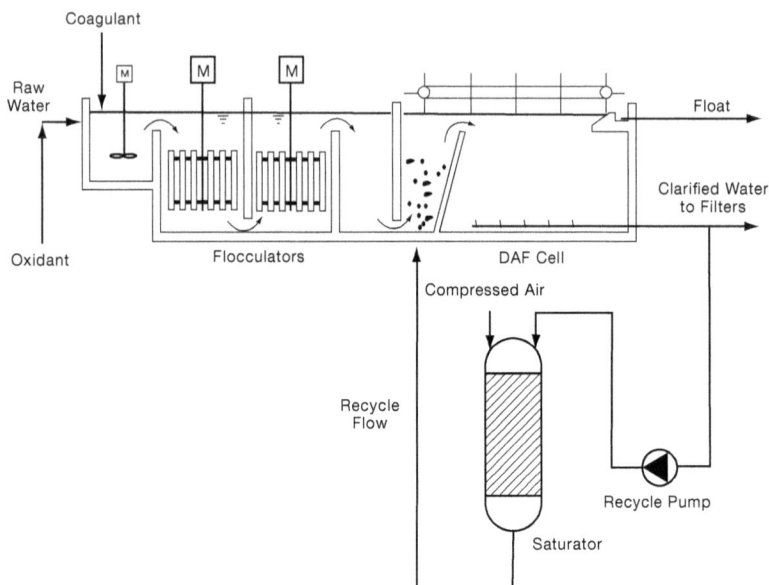

Figure 6-4   Typical dissolved air flotation process
Courtesy of Reid Crowther & Partners Ltd.

The process is rapid, much quicker than sedimentation, so DAF cells are substantially smaller than settling tanks of equal flow capacity. DAF systems are generally designed for basin loading of 6 to 10 gpm/ft^2, and loading rates of 15 gpm/ft^2 have been used. However, DAF works best with low-turbidity water that contains some organic material such as algae. As discussed in chapter 2, high-algae waters can be associated with reservoirs that also experience high manganese levels during turnover events. DAF is usually *not* the best process for treating river water with seasonally high silt loads.

Ballasted Flocculation

As discussed earlier, conventional oxidized iron and manganese floc particles settle at rate of approximately 0.63 in./min (1.6 cm/min), which is equivalent to a surface loading rate of approximately 0.5 gpm/ft^2. Plate and tube settlers can reduce the footprint of basin by as much as 75 percent and usually operate at a basin loading rate on the order of 2 gpm/ft^2.

Higher basin loading rates are possible by making the floc particle heavier through the addition of a microsand ballast. Basin loading rates on the order of 30 gpm/ft² are possible, and several full-scale plants treating a variety of surface waters have been in operation for more than 10 years. Figure 6-5 presents a schematic of ballasted flocculation system.

A ballasted flocculation system is well suited for sites with limited space. The system works well with high-turbidity waters and can respond quickly to changes in water quality. The system has a detention time generally of less than 15 min and can be placed into service very quickly. The settled water produced from a ballasted flocculation system is generally of very high quality, with settled turbidities generally less than 1.0 ntu.

Unlike the previously discussed clarification systems, which all generally produce sludge with a concentration of 0.2 percent to 1 percent, the sludge generated from a ballasted flocculation system is generally 0.05 percent to 0.4 percent. The thinner sludge generally requires more handling and treatment. Most of the ballast sand used in the process is recycled and reused via the hydrocyclone, but due to the sand recycling, additional maintenance is required, compared to more traditional clarification systems. The sand-recycle pumps are lined with hard rubber and the liners are a wear item with a 7- to 9-year life expectancy based on 24/7 operations. The bottom of the hydrocyclones (apex tip) is also a wear item, with a life expectancy of 2 to 3 years, and is easily replaced. Jar testing is recommended as an initial screening tool. If feasible, pilot testing is also recommended.

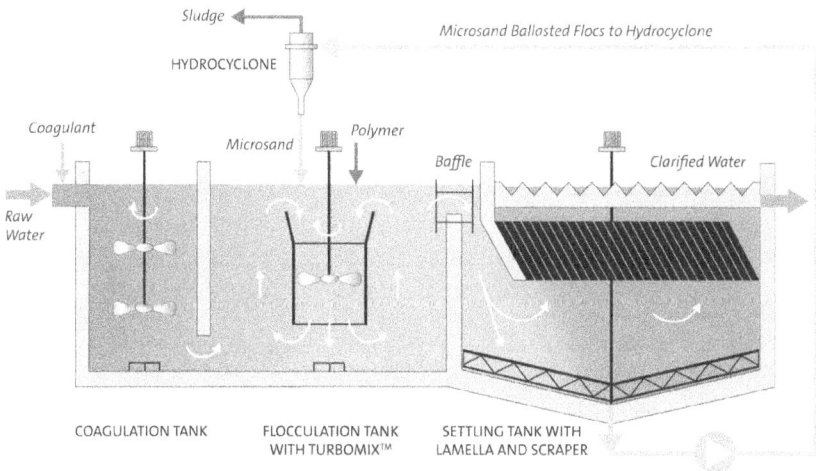

Figure 6-5   Ballasted flocculation system
Actiflo Turbo diagram courtesy of Kruger Inc.

SUSTAINABILITY

The sustainability issues associated with clarification include the following:

- Energy use
- Need for a superstructure
- Clarifier construction materials

The energy requirements of the clarification system are generally minimal, except for DAF, which requires a 6 percent to 10 percent recycle flow at a pressure of approximately 80 psi. Construction materials for the clarification basins and the equipment can be evaluated to minimize carbon footprint without compromising quality. A superstructure may be needed to prevent freezing and ice damage or to prevent algal growth on the basins.

The formation of a float/sludge layer on top of the basin usually requires that a superstructure be built for DAF operations.

If a superstructure is provided, the structure can be designed with sustainable features. The superstructure should allow ready access for maintenance while minimizing moisture levels in the building and reducing heat requirements. For example, in one midwestern city a superstructure was provided over tube settlers, but instead of leaving the entire upper portion of the tube settlers exposed, access points approximately 10 ft wide by 10 ft long were provided for inspection and housing of the tubes. This configuration greatly reduced the heat load of the building. Other sustainable features of the superstructure could include a green roof and solar panels.

7

Filtration

Significant advancements have been made in filtration for the removal of iron and manganese in recent years. Historically, filtration was limited to either conventional sand/anthracite or manganese greensand. (Manganese greensand and greensand refer to the same media and are used interchangeably.) Manganese greensand is made from glauconite, which is naturally occurring and is the substrate of manganese greensand. Glauconite requires hardening and coating, which involves the addition of chemicals such as alum and permanganate.

In response to the lack of availability of glauconite, the industry has developed alternative products. These products have several industry trade names, but can generally be classified into two categories: (1) manganese dioxide–coated sand and (2) manganese dioxide ore. Biological filtration and membrane filtration have also been used for the removal of iron and manganese. This chapter presents an overview of each of these filtration technologies, describes the types of waters that are suitable for each filtration system, and describes the general design and operating features for each filtration system.

FILTRATION SELECTION

The selection of the optimum filtration system depends on several factors, particularly the following:

- Source water (groundwater or surface water)
- Levels of iron and manganese
- Turbidity level of surface water
- Total organic carbon concentration of the raw water
- Additional contaminants in the raw water
- Residuals disposal limitations

Figure 7-1a presents a decision tree for groundwater systems and Figure 7-1b presents a decision tree for surface water systems. For groundwater systems it is important to know whether the system is under the direct influence of surface water (GUDI) as well as the concentration of iron and manganese. It is important to consider the following information when making decisions about treating groundwater with Figure 7-1a:

- If the combined Fe/Mn is >5 mg/L, clarification is needed with the gravity filtration option, but some membranes can treat a much higher Fe/Mn load.
- Ion exchange can remove hardness along with iron and manganese.
- If the source is GUDI, pressure filtration can be used in conjunction with a chlorine contact tank, cartridge filters, or UV, depending on requirements of the regulatory authority.

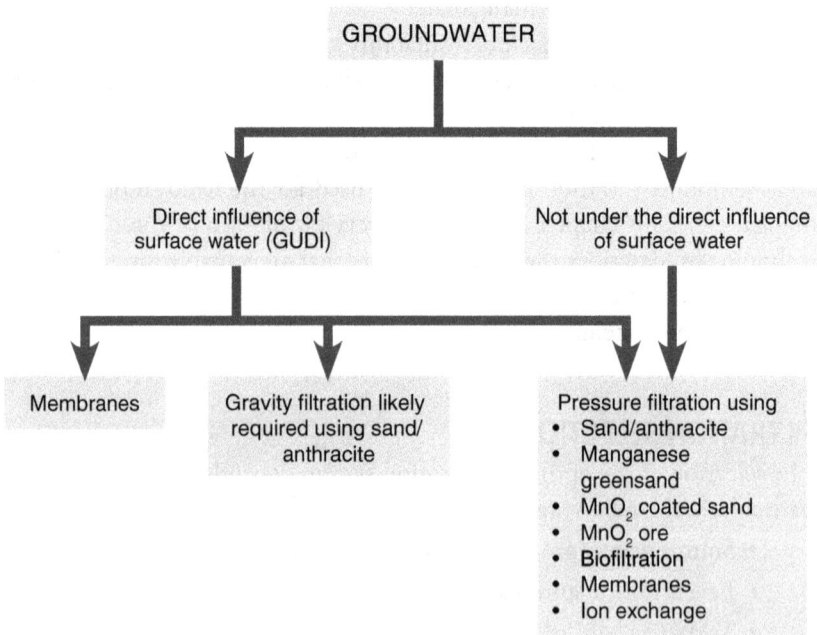

Figure 7-1a   Groundwater decision tree

SURFACE WATER

Low turbidity, low algae

High turbidity, high algae, high Fe/Mn, high TOC

Clarification

Membranes

Gravity filtration using sand/anthracite

Gravity filtration using sand/anthracite

Membranes

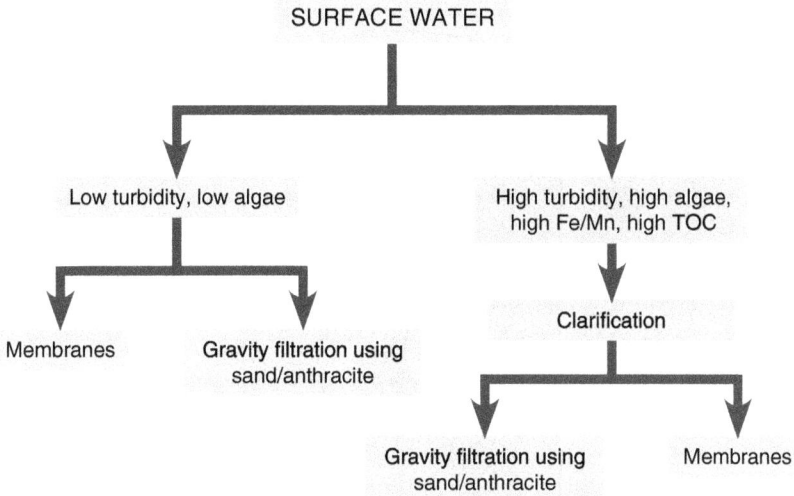

Figure 7-1b   Surface water decision tree

Pressure filtration using one of several different types of media is the most common form of treatment for the removal of iron and manganese from groundwater. Suitable pressure filtration media include greensand, manganese dioxide–coated sand, manganese dioxide ore, biofiltration, ion exchange, and sand/anthracite. For each of these systems a description of the media and its characteristics is provided in this chapter, followed by a description of the operating modes. Finally a design example is provided for the treatment of a 700-gpm well with 2 mg/L of iron and 0.5 mg/L of manganese.

Pressure filtration is accepted by most regulators for groundwater treatment, but it is generally not acceptable to regulators for surface water treatment because it does not remove pathogens. Flow distribution and backwashing are generally less efficient in a pressure filter than in a gravity filter. This inefficiency could potentially result in the passage of pathogens.

Pressure filtration for groundwater treatment is advantageous because it eliminates the need for a clearwell and the resultant need to double pump. Pressure filters are available in vertical and horizontal con-figurations. Vertical filters can generally treat up to 1 mgd as the diameter is limited to a maximum of 12 ft (Figure 7-2). Higher-capacity pressure fil-tration can be achieved using a horizontal configuration that can treat up to 5 mgd (12 ft in diameter by 40 ft long) (Figure 7-3).

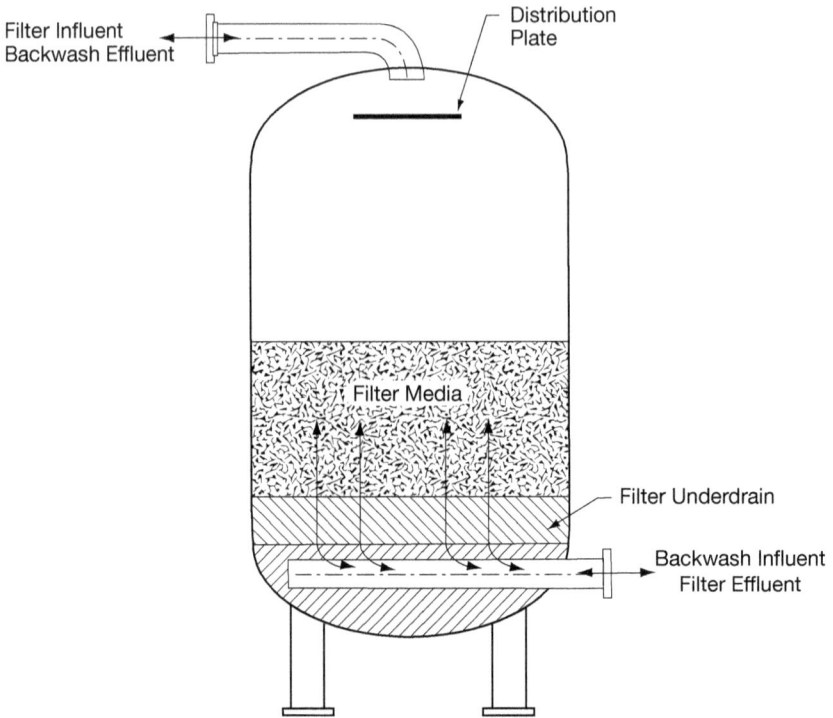

Figure 7-2   Vertical pressure filter
Courtesy of Anthratech Western Inc.

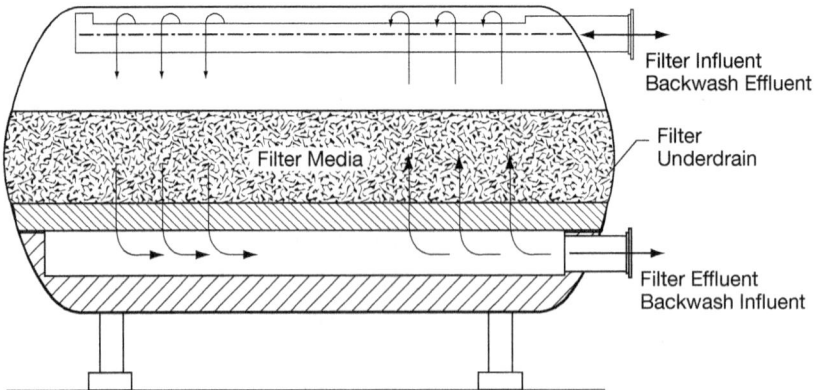

Figure 7-3   Horizontal pressure filter
Courtesy of Anthratech Western Inc.

In cold climates, vertical filters should be located inside a building because the piping connects to both the top and bottom of the vessel and is subject to freezing. Horizontal pressure filters can be located with the dished head inside the building and with the remainder of the unit located outside, since only the dished end contains the piping connections.

MANGANESE GREENSAND

The term *greensand* or *manganese greensand r*efers to a filter media derived from naturally occurring glauconite. The term *glauconite ore* refers to an alumina silicate that appears sand-like but is softer and similar to a clay material. New products that are not derived from glauconite, such as Greensand Plus (trade name of Inversand Company), have been introduced to the water treatment market in recent years.[1]

Manganese greensand has been used for several decades in North America, often specifically for iron and manganese removal. Manganese greensand is a purple-black granular filter medium processed from naturally occurring glauconite. The glauconite is synthetically coated with a thin layer of manganese dioxide [$MnO_2(s)$], and some particles have a definite green color, giving the material its common name. The only North American manufacturer of manganese greensand is located in New Jersey.

Glauconite exhibits an ion exchange capacity that allows the surface to be saturated with manganous ions. Following saturation, the glauconite is soaked in a strong oxidizing solution, which transforms the manganous ion to the insoluble $MnO_2(s)$ form. The surface coating makes up about 4.0 mg of a 1,000-mg greensand sample; about 0.4 percent of the weight of a particle of greensand is actually $MnO_2(s)$. Since $MnO_2(s)$ is black, particles not covered or only partially covered with the substance are easily identified by the greenish color of the underlying glauconite. Manganese greensand has an effective size of 0.30–0.35 mm, a uniformity coefficient of less than 1.60, and a specific gravity of approximately 2.4.

Glauconite has been used since the 1930s because it was inexpensive to coat with manganese dioxide and provided superior removal of iron and manganese. The systems were simple to operate and the media would last a minimum of 10 years. Some systems have been operating with original media for more than 50 years. Greensand systems are no longer available

---

[1] Raymond Jones, Hungerford & Terry, 2015. Personal communication.

and alternate media have been developed that exhibit similar properties. Some systems still have the original glauconite, and the following discussion presents operating modes and design information for these systems.

Conditioning

Virgin greensand (i.e., new and unused material) is not shipped in a regenerated form. Therefore, it must be conditioned prior to being put into service. The manufacturer recommends soaking the bed for not less than 4 hr in a solution containing about 3 oz (85 g/28,320 cm^3) of potassium permanganate ($KMnO_4$) per cubic foot (28,320 cm^3) of media. Greensand can also be regenerated using chlorine. Typically, a new bed is soaked in a solution containing about 2,000 mg/L chlorine for several hours.

Operating Modes

The typical operating mode for manganese greensand treatment involves preoxidation followed by filtration. In this sequence, called *continuous regeneration,* a strong oxidant such as $KMnO_4$ is added ahead of the greensand filter. In theory, by continuously preoxidizing both the iron and manganese to an insoluble particulate form, both the insoluble metals are removed by physical filtration/straining action through the greensand media. Any iron or manganese that is not oxidized (perhaps due to underfeeding of the oxidant) is *adsorbed* onto the $MnO_2(s)$ surface on the greensand particles.

Operating in this mode, a filter bed accumulates solids, which are then removed during the backwash cycle. The inherent stickiness of most oxidized iron and manganese compounds makes removal with backwash water alone a difficult task. Backwash assisted by air scour is recommended to keep the bed as clean as possible. Keeping the media clean is the key to a sustainable filtration process and extended media life. To prevent removing the medium's $MnO_2(s)$ coating by excessive air scouring, limit air scour to low rates, short periods, infrequent intervals, or a combination of the three.

A second operating mode for manganese greensand is oxidized iron filtration and manganese removal by adsorption. Typically, raw water is aerated to oxidize the iron, most of which is then removed in the coal layer that tops the greensand. The manganese in solution remains until it is adsorbed onto the fully regenerated $MnO_2(s)$ coating. This sequence is called *intermittent* regeneration. Intermittent regeneration is generally limited to waters with an iron level of less than 0.5 mg/L. Formulas have been

Table 7-1 Typical manganese greensand design criteria

Process Considerations	Manganese Dioxide–Coated Sand
Specific gravity of greensand	2.4
Surface loading rate	Up to 3 gpm/ft²
Backwash rate without air scour	12 gpm/ft², 7 min
Backwash rate with air scour	Not required for low levels of iron and manganese
Chemical feed	pH adjustment if pH is below 6.2 Chlorine
Bed depth	Anthracite cap 15–24 in. (38–61 cm) Greensand 15–24 in.
Effective media size	Anthracite cap (0.6–0.8 mm) Greensand (0.30–0.35 mm)
Capacity	1,000 grains/ft² of bed area
Manganese dioxide content by weight	0.4% to 1.0%
Media life	Approximately 10 yr

developed to calculate appropriate filter run lengths, but a practical guideline calls for backwashing and regeneration when the manganese level in the filter effluent reaches 0.05 mg/L. Optimum water quality occurs when filter runs end at some point before 0.05 mg/L manganese is reached in the filter effluent.

Table 7-1 presents a summary of the typical design criteria associated with a manganese greensand system.

Regeneration of Manganese Greensand

To continue removing iron and manganese in any operating mode, manganese greensand requires regeneration. Continuous regeneration is recommended for well waters where iron removal is the main objective with or without the presence of manganese. This method involves feeding sufficient $KMnO_4$ and/or Cl_2 to satisfy all chemical demands, including regeneration of sites on the $MnO_2(s)$ coating occupied by adsorbed iron and/or manganese.

Filter audits have demonstrated a need for careful dosing. Effluent from some filters receiving slightly pink water have contained higher levels of manganese than were found in the raw water. This effect resulted from $MnO_2(s)$ surfaces coated with oxidized iron, scale formation on the greensand particles, or a high percentage of manganese greensand particles backwashed away as the grains were replaced by heavier mudballs double the size of greensand (which have the appearance of manganese greensand when wet).

All three conditions can drastically reduce $KMnO_4$ demand. In other words, a slight pink feed provided too much regenerant for a filter bed with a very low regenerant demand. Because $KMnO_4$ is about 35 percent manganese, any overfeed of it may elevate manganese residuals in the finished water.

For Cl_2 dosing, the rule of thumb is to prefeed enough to satisfy all chemical demands and provide a free chlorine residual in the water leaving the treatment plant that is adequate to comply with regulatory standards.

These rules of thumb do not reflect a relationship between the regenerant used and the amount of iron and manganese accumulated in the bed. Some formulas take account of the molecular weights of the metals to be removed and the oxidants to be applied, allowing calculation of amounts of oxidant for specific quantities of water treated and metals removed, assuming a fully regenerated bed at the start of the filter run and the absence of other demands on the oxidant of choice. These formulas are best applied according to site-specific considerations derived from actual examples, because the results are then often weighted by other raw-water considerations or oxidant use upstream of the filter. Consult your treatment specialist or process consultant if these calculations are required.

Topping Coal

The use of a filter coal (anthracite) layer on top of a manganese greensand bed is often a critical step in reducing the solids load (i.e., filtering out particulate matter from pretreated water) and promoting optimum filter run lengths. The effective size, shape, and specific gravity of the topping coal are important choices. Poor combinations result in intermixing of the two media. A higher degree of intermixing reduces the capability of each medium to perform its intended function.

A filter coal layer over manganese greensand helps prevent damage to the greensand and helps with iron removal. The first benefit occurs because the coal particles are about three times larger than the greensand particles

(0.9 mm compared to 0.3 mm), which means that the spaces between the coal particles are also significantly greater and can hold a greater volume of filtered material than the manganese greensand can hold.

The space between particles of granular filter media is referred to as its *void area* (also called *bed porosity*). The amount of filtered material needed to fill these void areas is the filter medium's solids holding capacity. Coal can hold much more filtered material than manganese greensand before the flow of water becomes restricted. Typically, filter coal with an effective size of 0.7 to 0.8 mm has a void volume of approximately 60 percent.

Manganese greensand, on the other hand, can hold much less filtered material before the flow is restricted. Typically, the void volume of manganese greensand is 35 percent to 40 percent. Without a coal layer, the restricted flow through the manganese greensand bed can build up pressures high enough to fracture the manganese greensand particles and their $MnO_2(s)$ coating.

The void area in the coal also provides a place for oxidized iron and manganese to flocculate. When Cl_2 or $KMnO_4$ is added during pretreatment, the iron usually forms solid particles that join together into floc large enough to be caught between the particles of filter media. As the oxidized iron particles jostle through the coal layer, they collide with each other, stick together and to the coal particles, and are trapped, simplifying particulate removal.

If the coal layer did not prevent the oxidized iron from flowing down into the manganese greensand layer, the greensand would become coated with iron oxides over time, preventing further adsorption of manganese by the $MnO_2(s)$ and depleting the manganese greensand's ability to remove manganese.

Some exceptions complicate filtration. Some iron species under certain oxidation conditions remain fine enough to sift right through the filter bed. Some possible solutions to boost the particulate size are increasing detention time (i.e., the length of time the oxidizing chemical is in contact with the iron before the water reaches the filter), adding a flocculant (e.g., alum) following oxidation, and/or adding a polymer (sometimes referred to as a filter aid) before filtration.

Despite variations, a properly designed coal layer is an important tool for keeping the manganese greensand clean and extending its useful life. Proper design of the filter coal layer must ensure the optimum hydraulic backwash rate, since the rate for the filter coal is slightly less than the optimum rate for manganese greensand.

Because greensand is produced by only one manufacturer, operators can rely on a predictable consistency of particle size and specific gravity, which aids in calculations of appropriate backwash rates. The same is not true for topping coal. The specific gravity of filter coal can vary from 1.3 to 1.8, and uniformity coefficients can vary from 1.3 to 2.0, depending on the source of supply. An optimum design includes topping coal as coarse as possible that retains the ability to fluidize at a backwash rate just below the fluidization rate for the manganese greensand. To achieve this balance, coal with the right effective size, uniformity coefficient, and specific gravity must be chosen.

Regulatory/Design Guidelines and Design Example

Many regulatory codes limit the surface loading rate for filters to 3 gpm/ft^2 with a backwash rate of 10 gpm/ft^2. While most regulations do not specify the capacity of the manganese greensand, greensand generally has a capacity of 1,000 grains/ft^2.

For the 700-gpm design example with 2 mg/L of iron and 0.5 mg/L of manganese, the conceptual design would be as follows:

> Type of operation: continuous regeneration due to the presence of iron.
>
> Number of filters: 3 each at 350 gpm (2 duty/1 standby)
>
> Design loading rate: 3 gpm/ft^2
>
> Diameter per filter: $[(350 \text{ gpm}/3 \text{ gpm/ft}^2) \div \pi/4]^{1/2} = 12$ ft
>
> Area per filter: 113 ft^2
>
> Depth of anthracite: 18 in. (recommended by manufacturer)
>
> Depth of greensand: 18 in.
>
> Iron and manganese concentration (grains/gallon):
> (2 mg/L Fe + 0.5 mg/L Mn × 2) ÷ 17.1 mg/L/grains/gal
> = 0.17 grains/gal
>
> Capacity of greensand: 1,000 grains/ft^2
>
> Treated volume per filter:
> 113 ft^2 × 1,000 grains/ft^2 ÷ 0.17 grains/gal = 664,705 gal
>
> Run time per filter:
> 664,705 gal ÷ (350 gpm × 60 min/hr) = 31 hr
>
> Backwash volume:
> 12 gpm/ft^2 × 113 ft^2 × 10 min = 13,560 gal

Retrofitting Considerations

Greensand has a typical life span of approximately 10 years. After this period, the system will tend to exhibit the following characteristics:

- Filter run time will decrease and require more frequent backwashing.
- Chlorine and potassium permanganate doses will be higher.
- Head loss will be higher.

As greensand is no longer available, alternate media have been developed that are compatible with the existing vessel configuration and backwash systems. Manganese dioxide–coated sand can generally be done as direct replacement. Alternative media such as sand, manganese dioxide ore (pyrolusite), or Birm™ can be used as substitutes but require system modification. Coordination with the supplier is needed.

MANGANESE DIOXIDE–COATED SAND

As a result of the limited supply of greensand, manganese dioxide–coated sand has become available that exhibits properties similar to traditional greensand. The manganese dioxide–coated sand is available from Inversand, Clack, and others and is made by taking sand with an effective size of 0.3 mm and a uniformity coefficient of 1.6 and thermobonding a manganese dioxide coating. The filter media is 3 percent by weight manganese dioxide and has a specific gravity of approximately 2.5. The product has a high abrasion resistance and can be operated at higher loading rates than greensand.

The product was introduced in 2005, and most regulatory codes have not been revised to reflect the loading rates and backwash requirements. Most regulatory agencies allow design criteria to be established on a case-by-case basis and using data from similar installations. In some cases a pilot study may be required by the regulating agency.

Table 7-2 presents a summary of the general design and operating parameters associated with a manganese dioxide–coated sand system. A critical parameter that impacts the size and cost of the facility is the loading rate. The manufacturer has documented that a loading rate as high as 12 gpm/ft^2 will provide adequate removals. However, many regulators have allowed the use of 6 gpm/ft^2 without pilot testing. If a loading rate above 6 gpm/ft^2 is desired because of site or cost constraints, consideration should be given to performing a pilot test.

Table 7-2   Typical manganese dioxide–coated sand design criteria

Process Considerations	Manganese Dioxide–Coated Sand
Specific gravity of manganese dioxide–coated sand	2.4 to 2.6
Surface loading rate	Up to 12 gpm/ft^2; most regulators have allowed a loading rate of 6 gpm/ft^2
Backwash rate without air scour	12 gpm/ft^2, 7–10 min
Backwash rate with air scour	Not required for low levels of iron and manganese
Chemical feed	pH adjustment if pH is below 6.2 Chlorine
Bed depth	18 in. anthracite cap with an effective size of 0.6–0.8 mm 18 in. manganese dioxide–coated sand with an effective size of 0.30–0.35 mm
Capacity	900–1,200 grains/ft^2 of bed area
Manganese dioxide content by weight	3% Media is manganese dioxide–coated silica sand
Cost considerations	Anthracite: stable—multiple suppliers Maganese dioxide coated sand: stable—cost has not increased since media was introduced in 2005
Media life	More than 10 yr

Conditioning

Manganese dioxide–coated sand is conditioned in the field at startup by soaking the media in a 2 percent solution of potassium permanganate for approximately 4 hours. Approximately 0.25 lb of potassium permanganate is required per cubic foot of media.

Operating Modes

The typical operating mode for manganese dioxide–coated sand is what is referred to as catalytic oxidation, which entails continuously feeding of chlorine to oxidize the iron, manganese, and any other constituents such as hydrogen sulfide or ammonia that have a chlorine demand. Additional chlorine is added so the residual leaving the filters is between 0.5 and 1.0 mg/L. The additional chlorine serves to continuously regenerate the media.

Regeneration

The system is continuously regenerated with chlorine. Unlike greensand, which may require the addition of potassium permanganate for periodic regeneration, manganese dioxide–coated sand does not need to be regenerated with potassium permanganate. When the product was introduced, some owners were uncertain about the ability of chlorine to regenerate the media, so potassium permanganate systems were installed for regeneration. However, according to the media supplier, many of these permanganate systems were never used.

Regulatory/Design Guidelines and Design Example

For the 700-gpm design example with 2 mg/L of iron and 0.5 mg/L of manganese, the conceptual design would be as follows:

Type of operation: catalytic oxidation

Number of filters: 3 each 350 gpm (2 duty/1 standby)

Design loading rate: 6 gpm/ft^2

Diameter per filter:
$[(350 \text{ gpm}/6 \text{ gpm/ft}^2) \div \pi/4]^{1/2} = 8.6 \text{ ft (use 9 ft)}$

Area per filter: 63 ft^2

Depth of anthracite: 18 in.

Depth of manganese dioxide–coated sand media: 18 in.

Iron and manganese concentration (grains/gal):
$(2 \text{ mg/L Fe} + 0.5 \text{ mg/L Mn} \times 2) \div 17.1 \text{ mg/L/grains/gal} = 0.17 \text{ grains/gal}$

Capacity of manganese dioxide–coated sand: 900 grains/ft^2

Treated volume per filter:
$63 \text{ ft}^2 \times 900 \text{ grains/ft}^2 \div 0.17 \text{ grains/gal} = 333{,}529 \text{ gal}$

Run time per filter:
$333{,}529 \text{ gal} \div (350 \text{ gpm} \times 60 \text{ min/hr}) = 16 \text{ hr}$

Backwash volume:
$12 \text{ gpm/ft}^2 \times 63 \text{ ft}^2 \times 10 \text{ min} = 7{,}630 \text{ gal}$

Air scour: None required

MANGANESE DIOXIDE ORE (PYROLUSITE)

Pyrolusite is the mineral term for naturally occurring $MnO_2(s)$. Pyrolusite is produced from $MnO_2(s)$ ore in the United States, Australia, Brazil, and South Africa. The ore is crushed to specific sizes needed for potable water filtration processes. The resulting particles are solid pieces of $MnO_2(s)$, eliminating the need to develop a skin of $MnO_2(s)$ on each particle. Because pyrolusite has a specific gravity in the area of approximately 4.0 (compared with 2.4 for greensand and manganese dioxide–coated sand), air scour is necessary to keep the particles of pyrolusite scattered throughout the sand bed.

Operating Mode. A typical bed is a single medium layer of pyrolusite. Air scour of the bed helps to reduce the instantaneous backwash rate, depending on the process requirements. Pyrolusite bed depths range from 36 in. to 48 in., depending on the nature of the raw water and the treatment program. For a process with continuous regeneration, only a relatively small amount of pyrolusite is required to provide the adsorption sites needed for relatively small amounts of manganese in the raw water. Direct filtration through pyrolusite beds can be effective for groundwaters with iron levels up to 12 mg/L. Pilot testing is often the best way to determine pyrolusite requirements, including run time and liquid loading rates.

A solid $MnO_2(s)$ medium survives a vigorous backwash program including air scour without the worry of surface attrition. Any surface attrition of pyrolusite only exposes a fresh $MnO_2(s)$ surface. Typically, $KMnO_4$ is not used in a pyrolusite process, although it could be. The iron is removed in the media via filtration, and the manganese is then removed by adsorption. Where possible, sufficient chlorine is fed ahead of the filter to keep the pyrolusite in a regenerated condition.

Where formation of unwanted chlorine by-products is a concern, chlorine is used intermittently to regenerate the bed after backwashing. As with any filter medium, cleaning a pyrolusite bed thoroughly is important. Air scouring pyrolusite provides excellent results with no significant change to the particle size or shape.

General Information About Pyrolusite

Because of pyrolusite's high specific gravity, some plants cannot provide the necessary backwash flow to fluidize and adequately clean the bed. Simultaneous water backwash and air scour do the best job of cleaning the media. Many plants conduct air scour first, followed by hydraulic wash, to reduce the backwash rate required for the pyrolusite.

Table 7-3 presents a summary of the general design and operating parameters associated with a manganese dioxide ore system. The manufacturer has documented that a loading rate as high as 15 gpm/ft^2 will provide adequate removal. Coordination with the regulatory authority regarding the loading rate is recommended.

Table 7-3   Typical manganese dioxide ore (pyrolusite) design criteria

Process Considerations	Manganese Dioxide Ore
Specific gravity of manganese dioxide ore	4.0
Surface loading rate	Up to 15 gpm/ft^2
Backwash rate without air scour	25 gpm/ft^2, 10 min.
Backwash rate with air scour	3 scfm/ft^2 and 5 gpm/ft^2, 3–5 min 20 gpm/ft^2, 7 min
Chemical feed	pH adjustment if pH is below 6.0 Chlorine
Bed depth	36–48 in. manganese dioxide ore with an effective size of 0.3–0.5 mm
Capacity	300 grains/ft^3 of bed volume
Manganese dioxide content by weight	70%–80% Media is manganese dioxide ore (pyrolusite).
Cost considerations	Manganese dioxide ore has a greater cost than manganese dioxide–coated sand.
Media life	More than 10 yr

Regulatory/Design Guidelines and Design Example

For the 700-gpm design example with 2 mg/L of iron and 0.5 mg/L of manganese the conceptual design would be as follows:

> Type of operation: continuous regeneration due to the presence of iron
>
> Number of filters: 3 each 350 gpm (2 duty/1 standby)
>
> Design loading rate: 6 gpm/ft^2
>
> Diameter per filter:
> $[(350 \text{ gpm} \div 6 \text{ gpm/ft}^2) \div \pi/4]^{\frac{1}{2}} = 8.6$ ft (use 9.0 ft)
>
> Area per filter: 63 ft^2
>
> Depth of anthracite: None
>
> Depth of manganese dioxide ore: 36 in. (3 ft)
>
> Backwash volume:
>   5 gpm/ft^2 × 63 ft^2 × 3 min + 20 gpm/ft^2 × 63 ft^2 × 7 min = 9,765 gal
>
> Air scour:
>   3 scfm/ft^2 × 44 ft^2 = 132 scfm for 3 min

OTHER FORMS OF HYBRID MEDIA

Birm (Burgess Iron Removal Method)

The Burgess Iron Removal Method (Birm) is another filtration technology. Birm has been used mostly in point-of-use applications, with only a few applications in municipal, commercial, and industrial water treatment. (The Burgess Company was well-known in decades past as a maker of car batteries.)

Physical characteristics. The Birm media work through a $MnO_2(s)$ catalyst impregnating aluminum silicate sand. The base material is treated with manganous salts to near saturation, and the manganous ion is then oxidized to $MnO_2(s)$ with permanganate, a process similar to the one used to manufacture manganese greensand. While manganese greensand weighs 1,361 kg/m^3 (85 lb/ft^3), the Birm medium weighs only 560 to 640 kg/m^3 (35 to 40 lb/ft^3). It is also lighter than anthrasand and pyrolusite. Because of the medium's light weight, backwash rates require strict control, and the product's low specific gravity prevents the use of a coal cap.

Birm operating mode. The Birm process datasheet recommends air as the only oxidant to be used ahead of the filter. In most common ground-waters, dissolved iron and manganese occur in the divalent ferrous and manganous states due to the presence of free CO_2. The medium acts as a catalyst between the dissolved oxygen and the soluble iron and manganese compounds, enhancing this reaction and producing ferric and manganic hydroxides that precipitate to filterable form. After backwashing to remove the precipitate from the media, the filter is again ready for service flow. Acting as a catalyst, the Birm medium is not consumed and requires no regeneration.

General information about Birm. The main factor in the success of Birm is the operating recommendation to use air as the only oxidant. Experience clearly demonstrates that the adsorptive and oxidative capacity of the medium's $MnO_2(s)$ surface becomes exhausted without continuous or intermittent feeding of some regenerating oxidant. Although the media can be regenerated using Cl_2, a strong Cl_2 solution might affect any exposed surfaces of the base granular material. This concern arises from the manufacturer's caution against using free chlorine levels above 0.5 mg/L of chlorine.

Attrition studies support a prediction that vigorous air scour of the Birm filter would result in high attrition losses because of its low hardness factor. Because Birm has such a low density, air scouring is not necessary and not recommended.

Birm™ is a registered trademark of the Clack Corporation, located in Wisconsin. A specification sheet issued by Clack (1988) offers the following advice:

> Birm™ acts as an insoluble catalyst to enhance the reaction between dissolved oxygen (DO) and the iron compounds ..., [The reaction] produces ferric hydroxide which precipitates and may be easily filtered.... [The media] are easily cleaned by backwashing to remove the precipitate ... When using Birm™ for iron removal, it is necessary that the water contain no oil or hydrogen sulfide (H_2S can be removed by air stripping/aeration prior to the filter), organic matter (TOC) not exceed 4–5 ppm, the DO (i.e., dissolved oxygen) content equal at least 15% of the iron content with a pH of 6.8 or

> more… . A water having a low DO level may be pre-treated by aeration. Chlorination above 0.5 ppm free chlorine greatly reduces Birm's™ activity and dosages should be held at a minimum… Clack's Birm™ may also be used for manganese removal… . In these applications the water should have a pH of 8.0–9.0 for best results. If the water contains iron, the pH should be below 8.5. High pH conditions may cause the formation of colloidal iron which is very difficult to filter out.

Obviously, certain characteristics of raw water must be known before treatment using Birm. Levels of DO, TOC, H_2S, and pH are critical, because they greatly influence the effectiveness of the product.

Anthrasand

Anthrasand is another filtration medium similar to greensand. A base material of standard anthracite coal and silica sand, sized appropriately to the filtration service, is coated with a thin layer of $MnO_2(s)$. One basic difference between anthrasand and manganese greensand is the method of applying the $MnO_2(s)$ coating. Anthrasand is placed in the filter, where it is soaked in a manganous salt solution for a prescribed time, typically 24 hr, before $KMnO_4$ is added to oxidize the manganous ion to the $MnO_2(s)$ form. This process is referred to as in-situ-generated manganese dioxide.

Anthrasand operating mode. The suggested operating mode begins with dosing the raw water with $KMnO_4$ ahead of the filter to oxidize the iron and manganese, which is then removed by the filter. Any unoxidized iron and manganese is adsorbed onto the $MnO_2(s)$ surface. Sufficient oxidant is fed to keep the $MnO_2(s)$ surface in a regenerated condition.

Anthrasand offers inherent attrition resistance because of the hardness of the base anthracite coal and silica sand. The low clean-bed head loss associated with a range of media-effective designs is another advantage.

General information about anthrasand. Establishing a complete and fully oxidized $MnO_2(s)$ in situ depends on the base filter media and surface characteristics. Generally, anthracite coal and silica sand do not have ion exchange properties, nor are their surface areas increased by many pockets or caves. However, surface chemistry and roughness of the base media do impact coating adherence. The supplier of anthrasand has determined which base media supplies have improved coating adherence. Although many anthracite coal and silica sand samples have taken on complete $MnO_2(s)$ coatings naturally within filters on site, the time requirements of the process vary, taking years in some cases and only weeks in others.

Another mystery is why air scour seems to have little adverse effect on some sands naturally coated with $MnO_2(s)$, while air scour removes flakes of what appears to be $MnO_2(s)$ from sands in other filters. When naturally coated filter media are air scoured, closely monitor the backwash water to determine whether or not flaking is occurring.

A decision to use anthrasand, or any filter medium, is predicated on the nature and treatability of the raw water. Thorough pilot testing should be completed prior to the final design to allow evaluation and development of the most efficient possible removal process.

BIOFILTRATION

The biological filtration process operates hydraulically similarly to a pressure filter in that raw water is pumped through a pressure vessel containing a granular medium. However, unlike most other pressure filtration systems that rely on the formation of a chemical precipitate and subsequent filtration, biological processes do not require any chemical oxidants. Figure 7-4 presents a schematic of a biofiltration system.

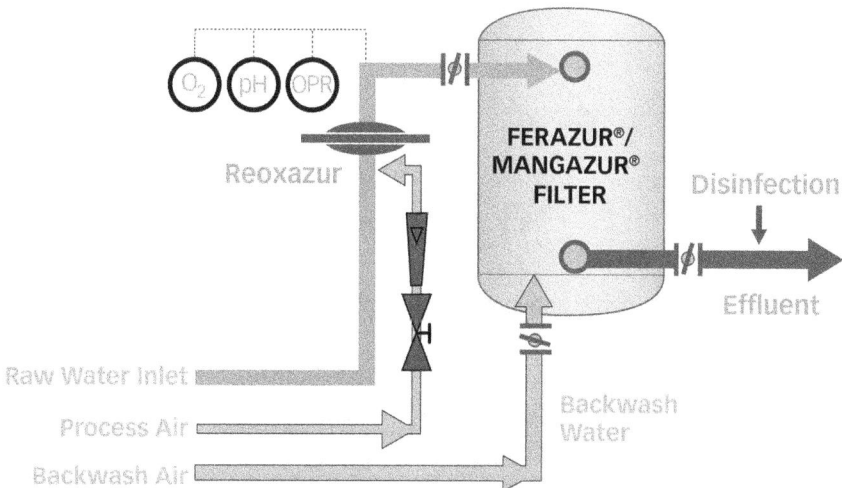

Figure 7-4   Schematic of biofiltration system
Courtesy of Infilco Degremont Inc.

Instead, conditions are established in the pressure vessel that foster the growth of bacteria. These bacteria oxidize the iron and manganese in the raw water, which is then retained within the filter in the form of dense precipitates. These precipitates are more compact than the amorphous precipitates formed during chemical oxidizing processes; therefore the biological filter has a higher iron and manganese retention capacity (up to five times higher). Figure 7-5 represents the relationship between the media, biofilm, and trapped solids.

The increased metal retention capacity allows the system to achieve long filter run times. Air is continuously injected into the raw water to provide the proper growth environment for the bacteria. It is important to note that the required environmental conditions for biological iron removal are different from those for biological manganese removal. Therefore, where both iron and manganese are present, two stages of biological filtration are required, one for biological removal of iron and one for biological removal of manganese. Biological filtration processes for removal of iron and manganese are proprietary patented systems manufactured by Infilco Degremont. They are marketed under the names Ferazur and Mangazur.[1]

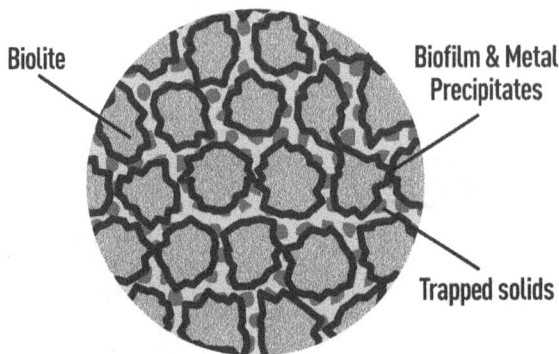

Figure 7-5   Relationship between media, biofilm, and trapped solids
Courtesy of Infilco Degremont Inc.

---

[1] Ryan Hess, Ondeo, 2015. Personal communication.

Biological treatment requires specific raw-water qualities and conditions, and not all groundwaters or surface waters can be treated economically using this technique. Where it can be used, biological treatment offers lower operating and capital costs than comparable physical/chemical processes, and it produces less waste product, which allows easier dewatering and disposal.

The ability of certain types of bacteria to absorb/adsorb dissolved iron and manganese, reducing the presence of metals by using enzymatic/catalytic action, has long been known. The biological treatment process encourages the growth and maintenance of large colonies of these bacteria, usually within a filter, where they can act on the dissolved iron and manganese. The bacterial action oxidizes the iron and manganese, and the resulting precipitates are trapped within the surrounding filter medium.

The process is generally rapid, with filtration rates substantially higher than those in physical/chemical treatment processes. Success depends on creating the right environmental conditions within the filter to permit the most beneficial bacteria to develop and maintain a strong colony. However, since different environmental conditions are required for each of the bacteria that remove iron and manganese, treatment of a water containing both elements requires a two-stage process. Usually, an initial biofiltration stage removes iron, and a second biofiltration stage removes manganese.

Biological Iron Removal

Many bacteria oxidize ferrous iron to the ferric form, causing it to precipitate. Bacteria commonly used in this role include *Gallionella*, *Leptothrix*, *Crenothrix*, and *Siderocapsa*. Generally the bacteria oxidize the iron using either of two processes:

- Intercellular oxidation (i.e., oxidation inside bacterial cells) by enzymatic action
- Extracellular oxidation (i.e., oxidation outside bacterial cells) by catalytic action of excreted polymers

As mentioned previously, desired bacteria require the proper conditions to develop. Figure 7-6 shows the field of activity of iron bacteria, using pH and redox potential as the graph axes.

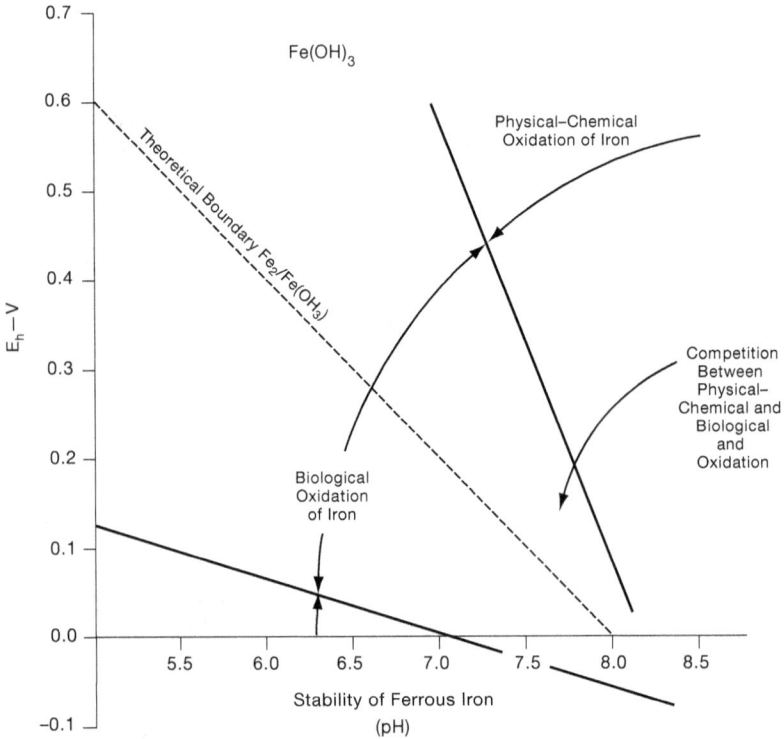

Below the 0.0 line, anaerobic (i.e., oxygen-free) conditions prevail; above the line, conditions are aerobic (i.e., oxygen is present). This graph indicates that the best rate of biological oxidation of Fe takes place at E_h-V 0.2 to 0.4. This environment provides just the right amount of oxygen. See the glossary under *redox* for the meaning of E_h-V.

Figure 7-6   Field of pH and redox potential activity of iron bacteria
Courtesy of Reid Crowther & Partners Ltd.

Generally, successful operation requires a pH level of 6.5 to 7.2, DO of 1 to 3 mg/L, and a water temperature of 10° to 25°C (50° to 77°F). These guidelines cover a broad range of conditions. Under some circumstances, pH levels in excess of 7.2 or below 6.5 do not hinder biological iron removal; in some instances, only very low levels of DO are required, and the temperature range may vary, depending on the type of iron bacteria used.

The raw water must not contain disruptive amounts of compounds toxic to the bacteria, however. Toxic compounds include the following:

- Chlorine (Cl₂). For this reason, chlorination for water disinfection should follow completion of the biological removal processes
- Hydrogen sulfide
- Heavy metals
- Ammonium nitrogen (NH₃)
- Phosphates
- Organics
- Hydrocarbons (i.e., any compound containing only hydrogen and carbon, such as benzene and methane)

In a typical biological iron removal process (Figure 7-7), the raw water is first oxidized using direct air injection, ensuring ideal redox potential

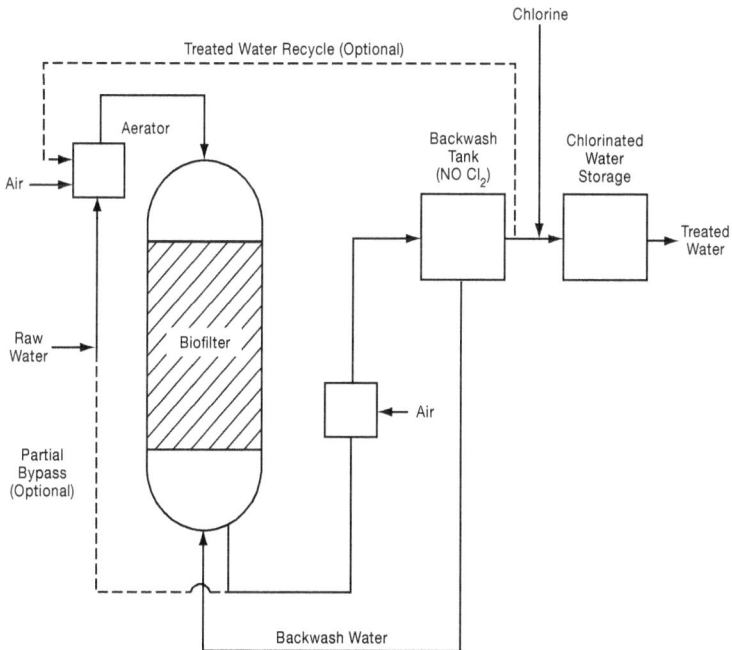

The raw water is first oxidized using direct air injection. This illustrates the importance of tight control over the amount of air fed, which should encourage bacteria to grow, without promoting physical oxidation of the Fe. This balance requires precise calibration and monitoring of the oxidation process. Also, note that backwsh water is held in a separate storage tank free of chlorine. Given an appropriate dosage and the correct amount of time, backwash water with a chlorine residual would kill all bacteria in the biofilter.

Figure 7-7   Typical biological iron removal process
Courtesy of Reid Crowther & Partners Ltd.

and DO level for bacterial growth. The process must avoid overoxidizing the raw water, or conditions appropriate for physical/chemical removal will develop, especially with pH greater than 7. This method requires precise calibration and monitoring of the oxidation process.

Bacteria can develop within both open and pressurized filters. The process forms a much more compact oxidized iron precipitate than normally found in physical/chemical removal processes. Consequently, high filter rates in the range of 20–60 m/hr (8–24 gpm/ft^2) can be used. These high rates call for a relatively coarse granular filter medium of about 1.2 to 2.0 mm effective size to limit head loss.

The filter is backwashed in the normal manner, but only with non-chlorinated backwash water. Backwash must also avoid excessive disturbance of the biofilm (i.e., the very thin layer of bacteria on each particle of the granular filter medium). Backwashing normally can use only water. Some designers have suggested that air scour should be avoided as the vigorous cleaning action tends to remove more biofilm than desirable from filter media, leading to reduced performance in subsequent filter runs. Normally, filters can be returned to service immediately following backwashing or after a prolonged shutdown without significant loss of performance.

When a new plant is started up for the first time, bacteria normally develop naturally, reaching adequate levels within two to three days in some cases and a week or so in others. Operators need not seed the filter with bacteria colonies. Other advantages of the biological process are the following:

- Ease of filtering biologically reduced iron
- High production rates
- Ease of dewatering sludge in the backwash water

Usually four to five times more water can be treated between backwashes using a biological process than using a conventional physical/chemical process. The precipitates of slightly hydrated iron oxides are much more compact than the precipitates formed in the physical/chemical removal processes (Voorinen et al. 1988). The dense sludge is less likely to clog filters and is easier to thicken and dewater. Backwash waste can also easily be thickened by gravity settling, and the resulting sludge can be dewatered by centrifuging (spinning it in a tub until all solids gather on the wall of the tub while the water escapes) or belt pressing (forcing the sludge between two belts that press out the water).

Biological Manganese Removal

Biological removal of manganese employs a process similar to that for biological iron removal. Both aerate the incoming raw water, both oxidize the metal (in this case manganese) through actions of bacteria, and both filter out the precipitates. However, the environmental conditions necessary to support the appropriate bacteria differ, so manganese removal must be done separately from biological iron removal.

Bacteria active in removing manganese are *Leptothrix*, *Crenothrix*, *Siderocapsa*, *Siderocystis*, and *Metallogenium*. Generally, the bacteria reduce the manganese through a biocatalytic process that includes intercellular oxidation by enzymatic action, adsorption of dissolved manganese at the surface of the cell membranes, and extracellular oxidation by catalytic action of excreted polymers. Manganese is deposited as manganese dioxide, $MnO_2(s)$, a black precipitate that is denser and more easily dewatered than manganese precipitates from a physical/chemical process.

The filter must support fully aerobic conditions (i.e., significant amounts of oxygen must be present). This condition demands much more vigorous aeration than for iron removal. Typical conditions needed within the filter are pH above 7.5, DO greater than 5 mg/L, and redox potential 300 to 400 mV.

The raw water must not contain materials toxic to the bacteria. The list of toxic substances and compounds is similar to that for biological iron removal. In particular, chlorine (Cl_2) in the raw water or the backwash water is fatal to the process.

Aeration is required upstream of the filter to ensure a DO level of greater than 5 mg/L within the biofilter. On-line air injection, spray aeration, or cascade aeration towers are common aeration methods. High filtration rates of 10 to 40 m/hr (4 to 16 gpm/ft^2) are possible for this method. A filter sand medium should have an effective size of 0.95 to 1.35 mm. If necessary, manganese removal can be enhanced by using a filter medium precoated with manganese dioxide, $MnO_2(s)$. Manganese greensand is too fine, with its effective size of 0.30 to 0.35 mm.

A new biological manganese removal plant requires a longer startup than is needed for biological iron removal. Development of sufficient numbers of bacteria organisms may take 2 to 8 weeks.

Biological Removal of Both Iron and Manganese

Removing both iron and manganese requires a two-stage process, as shown in Figure 7-8. The treatment process would include initial aeration and filtration for biological iron removal, secondary aeration to elevate the DO levels again, pH adjustment above 7.5 (using lime, soda ash, or caustic soda), secondary filtration for biological manganese removal, and finally some means of disinfection (such as chlorination).

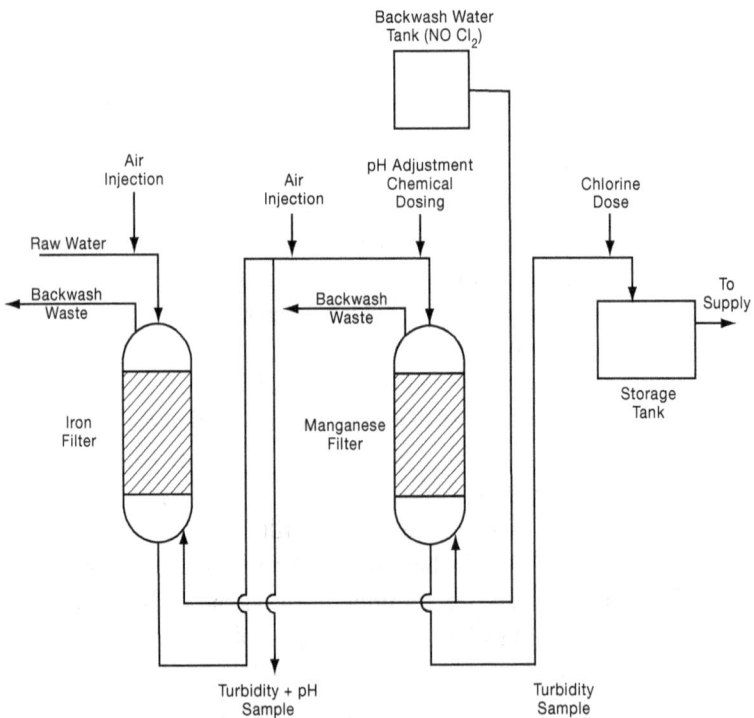

Biological processes remove Fe separately from Mn, since dissolved oxygen levels and redox potentials are different for each. (Note the two air injection locations.) Although Fe can be removed at a higher filtration rate (i.e., 20–60 m/h) than can Mn (i.e., 10–40 m/h), the slower Mn removal filtration rate must be used in a continuous removal process of the type illustrated here. Note also the separate backwash storage tank, in which the water is kept free of any chlorine.

Figure 7-8   Biological removal of iron and manganese
Courtesy of Reid Crowther & Partners Ltd.

Table 7-4   Typical biological iron and manganese filtration design criteria

Process Considerations	Biological Iron Filtration	Biological Manganese Filtration
Specific gravity of media	Varies	Varies
Surface loading rate	Up to 20 gpm/ft^2	Up to 15 gpm/ft^2
Backwash rate without air scour	6–10 gpm/ft^2	6–10 gpm/ft^2
Backwash rate with air scour	4–6 gpm/ft^2	4–6 gpm/ft^2
Chemical feed	Possible pH adjustment, but rare	Possible pH adjustment, but rare
Bed depth	3–6 ft	3–6 ft
Media life	10–15 yr	10–15 yr

Biological manganese removal is not practiced as frequently as biological iron removal. Where both iron and manganese appear in the raw water, biological iron removal is sometimes followed by chemical treatment to remove manganese by oxidation or adsorption. Published sources offer additional information about biological iron and manganese removal (e.g., Mouchet 1992).

Table 7-4 presents a summary of the general design and operating parameters associated with a biological iron and manganese filtration system.

Regulatory/Design Guidelines and Design Example

For the 700-gpm design example with 2 mg/L of iron and 0.5 mg/L of manganese the conceptual design would be as follows:

> Number of filters: 2 for iron (700 gpm each) and 2 for manganese (700 gpm each)
>
> Design loading rate for iron removal: 12 gpm/ft^2
>
> Design loading rate for manganese removal: 8 gpm/ft^2
>
> Iron filter [(700 gpm ÷ 12 gpm/ ft^2) ÷ (π/4)]$^{1/2}$ = 8.6 ft (use 9 ft)
>
> Area of iron filter: 63 ft^2
>
> Manganese filter: [(700 gpm ÷ 8 gpm/ft^2) ÷ π/4]$^{1/2}$ = 10.6 ft (use 11 ft)

Area of manganese filter: 95 ft^2

Depth of media: 48 to 60 in. (suitable for both iron and
manganese filters)

Information on filter run time and backwash volume is determined on a case-by-case basis and should be obtained from the supplier.

ION EXCHANGE

Ion exchange is a traditional method to remove calcium and magnesium, which are both divalent metal cations. Calcium and magnesium are referred to as hardness and the ion exchange system is referred to as a softener.

The softener operates by exchanging dissolved ions in the water with ions that are impregnated on a zeolite medium. Because iron and manganese are both dissolved cations, a water softener will also remove up to several milligrams per liter of iron and manganese. A typical process is shown in Figure 7-9. The flow to be softened is pumped into a pressure filter containing a chosen zeolite media, which removes hardness as well as free cations of iron and manganese.

Two types of cation exchange media are used as part of the removal of iron and manganese: strong acid and weak acid. Strong acid media exchange *sodium* for the calcium, magnesium, iron, and manganese. In general, 2 mg/L of sodium is added to the water for every 1 mg/L of iron or manganese removed. Strong acid media are regenerated with sodium chloride.

Weak acid media exchange *hydrogen* for the calcium, magnesium, iron, and manganese. The hydrogen exchange decreases the pH of the water. Weak acid media are regenerated with hydrochloric acid.

Following regeneration, the zeolite medium is rinsed with clean water before being returned to service. An important restriction requires that no oxidants be added to the water on its way to the softener. Otherwise, chemical oxidation of the iron and manganese may occur, plugging the zeolite medium or coating it with oxidation products.

Ion exchange is often applied for point-of-use treatment. However, sodium cycle systems are being used less frequently due to potential concerns with the addition of sodium to the filtered water and the generation of a brine waste that presents disposal concerns. Weak acid systems are a viable option for iron and manganese treatment in instances where hardness removal is needed and technologies such as nanofiltration and reverse osmosis are not feasible or cost effective.

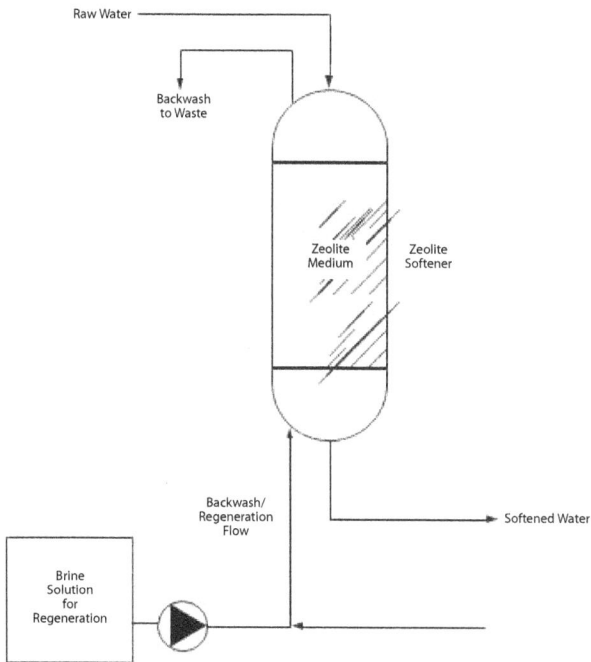

As water flows through the zeolite medium, Fe and Mn are held by ion exchange until the zeolite's capacity has been reached. Regeneration using a brine solution exchanges the held Fe and Mn ions for sodium ions, and the Fe and Mn is then backwashed away to waste.

Figure 7-9   Zeolite softening schematic
Source: Reid Crowther & Partners Ltd.

MEMBRANES

Hollow-fiber membranes are another option that can cost-effectively remove iron and manganese. Hollow-fiber membranes operate at pressures of 4 to 20 psi and have been used in the treatment of drinking water for more than 20 years. As described in Figures 7-1a and 7-1b, hollow-fiber membranes merit consideration in different situations, including for:

- Well water considered to be groundwater under the direct influence of surface water (GUDI)
- Well water with iron and manganese above 5 mg/L
- Surface water where membranes are being considered as an alternative to sand/anthracite filtration

Groundwater Under Direct Influence with Iron and Manganese

Consider the source water with 2 mg/L of iron and 0.5 mg/L of manganese in the Regulatory/Design Guidelines and Design Examples in this chapter. If this source is found to be under the direct influence of surface water, compliance with the requirements of the Surface Water Treatment Rule will be necessary. This means that advanced filtration to enable the removal of *Giardia* and *Cryptosporidium* will be needed.

Traditional iron removal media are generally not designed for pathogen removal, and an additional filtration system is needed to remove pathogens such as *Giardia* and *Cryptosporidium*. Modifications to traditional systems such as coagulant addition may be feasible, but these modifications may compromise the system performance.

One option is hollow-fiber membrane filters that are specifically designed for *Giardia* and *Cryptosporidium* removal without the addition of a coagulant. These membranes, which have a pore size of about 0.1 microns, strain out the pathogens, which are 3 to 5 microns in size. The dissolved iron and manganese that is not organically complexed can easily be oxidized, and the resulting particulate iron and manganese can be removed on a membrane filter. The membrane filtration option eliminates the need for a dedicated pathogen filtration system and coagulant addition.

Hollow-fiber membranes are also advantageous when the source water contains high levels of metals. If the combined iron and manganese is above 5 mg/L, traditional filter run times may be less than 24 hr and large volumes of filter backwash will be generated. In these cases, clarification is often provided upstream of the traditional media system. Some hollow-fiber membrane systems can handle up to 50 mg/L of combined iron and manganese operating in direct filtration mode and achieve a 90 percent recovery. The use of the membrane system saves space and eliminates the need for a clarification system.

Membrane Removal of Manganese in Surface Water

Sand/anthracite filters are traditionally used in surface water treatment. The filters are often located above a clearwell. Chlorine addition to the top of the filters is a convenient way of adding chlorine to provide pathogen inactivation within the clearwell.

An unintended benefit of the chlorine addition is the creation of a manganese dioxide coating on the sand particles that allows for the removal of manganese. The sand particles behave similarly to the manganese dioxide–coated media described earlier, in that the coating is continuously regenerated by the addition of chlorine.

The manganese in surface water is often organically complexed and behaves differently than manganese in groundwater. Newer water treatment plants are often designed with membranes instead of sand/anthracite filters. However, membranes do not have the same manganese-removal properties as a manganese dioxide–coated sand, even with the addition of chlorine prior to the membranes, and it is possible that manganese would not be completely removed. If membranes are being considered for a source water with elevated manganese, identifying the manganese in the source water and running an extended pilot test are recommended.

SUSTAINABILITY

For the treatment of a specific water, each filtration option is different in the following ways:

- Equipment size
- Chemical usage
- Energy usage
- Residuals volume, concentration, and characteristics
- Media life

Traditionally, life-cycle costs are used to compare filtration options. However, a sustainability analysis of each option may assist in selecting the optimum filtration option. The methodology described in chapter 2 provides a framework for assessing the sustainability of each option.

8

Residuals

The iron and manganese treatment processes described in the previous chapters generate wastes that require disposal. The wastes generated from the treatment processes are referred to as *residuals*.

Regardless of the type of unit processes used for removing iron and manganese from the source water, additional residuals-handling systems are necessary. These systems can range from the most basic, such as lagoons or conveyance to a sanitary sewer, to more elaborate dewatering systems.

Historically, residuals from iron and manganese facilities were disposed of in local watercourses or stored in lagoons with limited treatment. Now, the public and regulatory communities require that residuals be handled sustainably in order to minimize the production of wastes from the treatment processes. In addition, there is a focus on reducing the energy and chemicals used in the treatment of residuals, as well as developing alternative beneficial uses for the wastes.

Several handbooks are devoted to the design and operation of residuals handling systems for drinking water treatment plants, and these are listed in the References under "Additional Resources." The purpose of this chapter is to describe important aspects of residuals handling related specifically to iron and manganese treatment systems.

The first section discusses source water quality parameters that have a significant impact on the generation of residuals. Next, the residuals impacts on the oxidation, clarification, and filtration processes are discussed. Local issues such as land availability, the capacity of the local sewer and wastewater treatment facility, and regulations are addressed because local conditions provide the boundary conditions for establishing residuals-handling options. An overview of residuals handling and treatment options is presented, followed by a design example using the source water described in chapter 7. This chapter concludes with a discussion of important sustainability considerations.

SOURCE WATER QUALITY

Source waters that contain iron and manganese above the regulatory limits sometimes contain other elemental contaminants, including radium, radon, arsenic, and hardness. Laboratory detection methods have become more precise, and the industry is now able to detect these contaminants in smaller quantities. If these contaminants are below the maximum contaminant level, specific treatment is not needed. However, the traditional treatment methods for iron and manganese often remove many of these contaminants, which become a part of the residuals. During the residuals-handling processes, these contaminants can become more concentrated and influence the method of residuals handling.

Radium 226 and Radium 228

Radium 226 and 228 are radioactive. The USEPA MCL for combined radium 226 and 228 is 5 pCi/L. Manganese dioxide–type filtration media will remove a portion of the radium. Some of the radium is removed in the backwashing process, and some is retained on the media. The backwash waste may have elevated levels of radioactivity that may affect residuals handling. Furthermore, the portion of the radium that is retained on the media may increase the radioactivity of the filtration media, shortening its life and ultimately requiring removal and disposal. The presence of radium may also require special disposal procedures for the media.

Radon 222

Radon 222 is a radioactive isotope that can be found with uranium and radium. Radon exists as a dissolved gas in groundwater. Radon is not currently regulated in drinking water but a proposed regulation was issued in 1999. Radon is removed during any aeration process used for oxidation and discharged into the off gas from the aerator. If an aerator is being used for oxidation and radon is present, state regulators should be consulted regarding any special precautions or permits required.

Arsenic

The USEPA MCL for arsenic is 0.010 mg/L (10 ppb). Arsenic, if present in the source water, is often removed in the clarification and filtration processes used for the removal of iron and manganese. Arsenic will therefore be a component of the residuals and its concentration will increase in the residuals handling processes. If arsenic is present in the source water even at a concentration below the MCL, the residuals should be analyzed for arsenic.

Hardness

Hardness is regulated as a secondary standard with a USEPA SMCL of 250 mg/L. As discussed in chapter 6, systems with elevated iron and manganese often use a solids-contact clarifier. The pH is often increased to above 8 to increase the rate of oxidation of manganese.

For larger plants, lime addition is generally more cost effective than sodium hydroxide for raising the pH. When the pH is raised, some portion of the hardness will be removed in the solids-contact clarifier. The hardness removal will increase the quantity of residuals that are produced.

TREATMENT PROCESSES

The treatment processes for iron and manganese are oxidation, clarification, and filtration. Figure 8-1 describes the residuals that are generated from each unit process. The residuals generated from the clarifiers and filters are liquids, and if dewatering is provided, a solid *cake* is generated.

If aeration is used for oxidation and dissolved gases are present in the water, a certain portion of these dissolved gases will be removed and the air emissions are a form of residuals. The concentration of the dissolved gases in the raw water should be quantified and an estimate of their concentration in the aerator off gas should be computed to determine if there are any potential regulatory or environmental concerns.

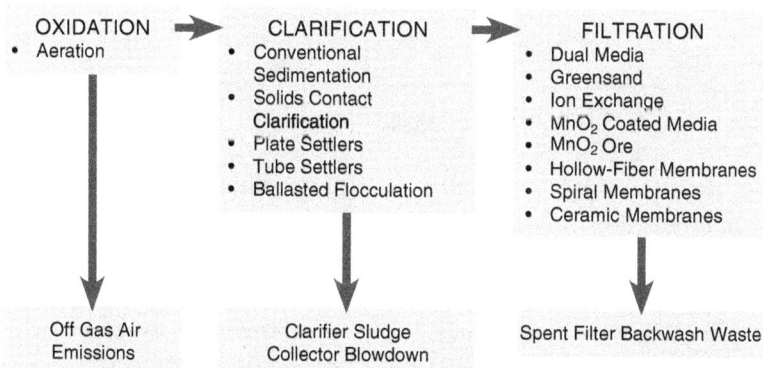

OXIDATION ➡ CLARIFICATION ➡ FILTRATION

- Aeration

- Conventional Sedimentation
- Solids Contact Clarification
- Plate Settlers
- Tube Settlers
- Ballasted Flocculation

- Dual Media
- Greensand
- Ion Exchange
- MnO_2 Coated Media
- MnO_2 Ore
- Hollow-Fiber Membranes
- Spiral Membranes
- Ceramic Membranes

Off Gas Air Emissions

Clarifier Sludge Collector Blowdown

Spent Filter Backwash Waste

Figure 8-1   Residuals generated from iron and manganese removal processes

The total amount of residuals generated by a water treatment plant on a daily basis can be estimated by using the following equation:

Dry solids of residuals (lb dry solids/day) =
Flow (mgd) × 8.34 lb/gal × Concentration of solids
removed (mg/L)

This represents the total dry-solids production for the plant. The clarifier typically generates approximately 90 percent of this total dry-solids waste production, and the filters generate the remaining 10 percent. The concentration of solids removed in the clarifier consists of the oxidized iron and manganese solids along with additional substances, such as hardness, that are removed. These additional substances for groundwater could include hardness, arsenic, radionuclides, and any other matter that is removed in the clarifiers and filters. The additional substances associated with surface waters would include turbidity and coagulated organics along with any dissolved substances that are converted to particulate form in the oxidation and coagulation processes. The residuals generation for the removal of iron and manganese can be calculated thusly:

1 mg/L of iron will oxidize into 1.91 mg/L of oxidized iron residual solids

1 mg/L of manganese will produce 1.61 mg/L of oxidized manganese residual solids

For example, consider a plant using chlorine for preoxidation with a flow rate of 2 mgd, with 3 mg/L of iron and 1 mg/L of manganese. The daily solids production from the clarifiers is computed as follows:

2 mgd × 8.34 lb/gal × (3 mg/L of iron × 1.91 + 1 mg/L
of manganese × 1.61) = 2 × 8.34 × (5.73 + 1.61) = 16.88
× 7.34 = 122 lb dry solids per day

Total dry solids production per day would be 122 lb. The clarifier sludge production would be 0.9 × 122 lb dry solids per day = 110 lb dry solids per day.

The total daily dry-solids production from the clarifier provides the basis for subsequent calculations. In the actual treatment process, clarification residuals are generated from the operation of the clarifier sludge collection system. The frequency of operation of the sludge collectors and the concentration depend on the number of clarifiers, the solids loading to the clarifier, and the type of clarifier.

Conventional Sedimentation Basins and Plate Settlers

Sludge removal from conventional sedimentation basins can be manual and performed several times per year, or mechanical, performed daily using sludge collection systems including chain-and-flight, perforated piping, and scrapers. Chain and flight collectors can handle sludge concentrations up to 0.5 percent, whereas perforated piping systems generally operate better with concentrations of 0.1 to 0.3 percent. Scraper systems can generally handle 1 to 2 percent solids. A mechanical sludge collection system is typically operated two or six times per day depending on the sludge generation rate.

Plate settlers are always equipped with automatic sludge collection systems, but in order to access the sludge collectors for maintenance, the plates may need to be removed. When choosing sludge collection systems, allow for sufficient clearance to access the operational controls and to maintain the collectors.

Solids-Contact Clarifiers

Solids-contact clarifiers are equipped with circular sludge collection systems and operated with a sludge blanket with a solids concentration of around 0.5 percent. The clarifier's scraper mechanism is in continuous operation, moving the sludge to the center of the basin. A valve is then opened to withdraw the residuals.

Ballasted Flocculation

Residuals from a ballasted flocculation system are generated from hydrocyclones, which separate the sand from the residuals removed in the clarifier. The residuals generated from the ballasted flocculation system are approximately 0.05 percent to 0.04 percent and are therefore much lower in concentration than the residuals from other clarification processes. The lower concentration results in a higher volume of waste.

Filtration

Filtration systems generate residuals from backwashing the filters. This waste is often referred to as *spent filter backwash* or *backwash wastewater*.

The frequency and quantity of spent filter backwash depend on the type of filtration system. Chapter 7 provides an overview of backwash requirements for each type of filter. The volume of waste generated from the backwash of a single filter can be estimated in this way:

> Surface area of filter media in one filter (ft²) × backwash loading rate (gpm/ft²) × duration of backwash = gal/backwash

The frequency of backwash can be determined from the filter run time calculations described in chapter 7. Generally the concentration of the spent filter backwash ranges between 100 and 500 mg/L.

LOCAL AND REGULATORY ISSUES

Feasible residual-handling options are dictated by local conditions and can be determined by asking the following questions:

- Is a sanitary sewer located near the site, and does the sanitary sewer have enough sewer collection capacity to allow for residuals disposal via sewer?
- Does the wastewater treatment plant have the capacity to accept residuals?
- If a sanitary sewer is not available, is it feasible to truck liquid residuals to a local wastewater treatment plant?
- Is there land available to accommodate sand drying beds or lagoons?
- Is there a stream nearby that could accept treated liquid residuals, and can discharge water quality requirements be met?
- Are there landfills in the vicinity that could accept dewatered cake?

The responses to these questions will establish a subset of boundary conditions for residuals treatment options to be identified and costs to be estimated. For example, if a sanitary sewer is available with sufficient capacity but the wastewater treatment plant does not have capacity to accept the residuals, it may be possible to work with the local sewage authority to expand the plant. If the site is very small and local haulers are available, hauling of liquid residuals should be considered.

Regulations

Regulations for residuals handling include federal, state, and local requirements. At the outset of a project, the state regulatory agency should be consulted to review permitting issues and provide guidance on which residuals
handling systems have the most environmental benefit.

Despite the numerous regulatory requirements, some elements are
common among federal, state or provincial, and local agencies. Many
requirements related to residuals discharge come under the National Pollution Discharge Elimination System (NPDES) regulations and are often
adopted and administered at the state and local levels.

If liquid residuals are discharged via a sewer to a wastewater treatment
plant, the regulations generally include compliance with the requirements
of an industrial pretreatment program. Depending on the quantity and
quality of the residuals, a permit may be required from the area sewerage
authority. If liquid residuals are hauled off site for discharge at a wastewater
treatment facility, similar requirements by the receiving facility may apply.

If sand drying beds or lagoons are used with the intent of discharging
a portion of the liquid waste to groundwater, then a discharge to groundwater permit would likely be needed. The permit would stipulate the discharge water quality and monitoring requirements, as well as the local
subsurface characteristics. If sand drying beds or lagoons discharge to the
surface water, an NPDES permit or equivalent for discharge to surface
water would be required.

If residuals are dewatered, disposal options for the cake include disposal at landfills and beneficial reuse. Regulatory requirements concerning
landfills and beneficial reuse would come under 40 CFR Part 503 of the
US federal regulations.

RESIDUALS-HANDLING OPTIONS

Once the residuals quantities have been established and the local and regulatory boundary conditions identified, the residuals treatment systems can
be evaluated. The following is a list of potential residuals-handing options
that can be used for treatment of clarifier and filtration residuals.

- Direct sewer discharge. The wastewater treatment plant is the end
  point for disposal.

- Equalization and sewer discharge. Liquids and solids are combined, and the wastewater treatment plant is the end point for disposal.
- Thickening. Decant is recycled and thickened sludge is disposed of at a wastewater treatment plant, hauled off site, or dewatered.
- Clarification using batch or continuous processes. Decant is recycled or discharged to a local water body. The settled sludge is hauled off site, conveyed to the sewer, or dewatered.
- Nonmechanical dewatering. Liquid is recycled or conveyed to groundwater. Solids require disposal.
- Mechanical dewatering. Decant is recycled and cake is generated and hauled to a landfill or used for beneficial reuse.

Direct Sewer Discharge

Iron and manganese residuals are generally not biologically active. In most instances the chemical composition does not adversely affect the unit processes of a wastewater treatment plant. Typically the wastewater treatment plant is concerned with the volume and mass of residuals that could overload the unit processes. The dry-solids production is a function of the raw water solids and the oxidation chemistry, and little can be done to significantly reduce the dry-solids production. However, the volume of residuals can be minimized by clarification and thickening processes.

The residuals need to be conveyed to the sewer, and generally the velocity in the sewer should be a minimum of 2 ft/s to minimize residuals settling in the piping. Clarifier residuals with a concentration of 0.1 to 0.5 percent should not be problematic for gravity sewers, but care should be taken to evaluate the potential for accumulation of solids.

Spent filter backwash, which has a concentration of approximately 0.01 percent to 0.05 percent, does not tend to cause solids buildup within a sanitary sewer. Spent filter backwash typically has a high instantaneous flow rate, which may preclude direct sanitary sewer discharge. For example, a 10-ft-diameter vertical-pressure filter backwashed at 12 gpm/ft^2 generates an instantaneous flow rate of 940 gpm of spent filter backwash, and if backwashed for 10 min generates approximately 9,400 gal of wastewater. If the sanitary sewer does not have sufficient capacity for an instantaneous flow rate of 940 gpm, it may be feasible to store the spent filter backwash (i.e., equalize the residuals) and then discharge the residuals to the sewer at a lower flow rate.

Disposal to the sanitary sewer or the local publicly owned treatment works (POTW) is generally the least complex option, with the only equipment being pumps (if needed) and valves. The cost of disposal to most sewer systems is traditionally based on flow, biochemical oxygen demand (BOD), and total suspended solids (TSS). An initial, potentially significant, connection fee based on projected flow could also be assessed.

The allowable flow depends on available treatment capacity at the POTW. BOD concentrations of typical iron and manganese removal residuals range between 110 and 400 mg/L with an average of 220 mg/L, and the TSS concentrations range from 100 to 350 mg/L with an average of 220 mg/L (Metcalf & Eddy 2013).

POTWs typically establish local limits for BOD and TSS concentrations, and discharges with higher concentrations may be subject to surcharges. POTWs also establish local limits for other parameters such as oil and grease, pH, metals (such as iron, manganese, and arsenic), total dissolved solids, and other pollutants. The POTW may also require the discharger to periodically sample, analyze, and report on wastewater discharges. The local POTW should be contacted early on in the project to confirm available treatment capacity, local limits, and any permitting requirements for discharge to the sanitary sewer. Pretreatment may be required in some cases.

Cost for wastewater discharges with BOD and TSS concentrations below the local limits can range from $1.00 to $3.00 per 100 ft^3. BOD and TSS surcharges are highly variable, but can range from $0.50 to more than $1.00 for BOD and $0.25 to more than $1.00 for TSS per pound. Some POTWs may also have surcharges for ammonia and phosphorus.

Equalization and Sewer Discharge

Both the clarifiers' blowdown and spent filter backwash are generated intermittently. If the sewer system does not have the capacity to handle the instantaneous discharge from either or both of these residuals streams, it may be possible to equalize the residuals and then convey the residuals at a lower and constant rate to the sewer system. Consider a pressure filter system with three vertical pressure filters each 10 ft in diameter and with a 12-gpm/ft^2 backwash rate for 10 min. Each filter will have an instantaneous backwash rate of 940 gpm and generate 9,400 gal of backwash.

If the filter run time is 24 hr, approximately 28,200 gpd will be generated. If this volume is equalized and discharged over a 24-hr period, the rate would be 20 gpm. If equalization is provided, settling will likely occur in the equalization tank. To prevent settling, submersible or vertical mixers could be installed in the tank.

Thickening for Clarifier Residuals

The volume of the clarifier sludge can be reduced by thickening, which is essentially a form of sedimentation but with dual goals of reducing the volume of residuals and producing higher-quality decant water.

Thickeners can increase the solids concentration to 2 percent to 4 percent and produce decant water with a turbidity of less than 10 ntu. Thickeners are available in both gravity and mechanical forms. Gravity thickeners are generally larger than mechanical thickeners but are simpler to operate and allow for sludge storage in the bottom of the tank.

Mechanical thickeners include devices such as gravity belt thickeners, rotary drum thickeners, and centrifuges. Mechanical thickeners are much smaller than gravity thickeners, have more moving parts, and require some form of downstream sludge storage. Thickened solids are generally not appropriate for disposal to sanitary sewer piping systems, but they are especially suitable for hauling off site in a tanker truck. The 2 percent to 4 percent waste can then be disposed of to a wastewater treatment plant's thickener or to a local waste processing facility. The cost of hauling and disposal of thickened sludge varies but is generally on the order of $0.05 to $0.15 per gal. This cost typically includes both trucking and disposal costs.

The use of a thickener is usually required as part of any mechanical dewatering process and can improve the performance of a sand drying bed.

Clarification for Filter Residuals

If the spent filter backwash residuals cannot be disposed of in the sewer, even with equalization, treatment can be provided in the form of clarification. The clarification process will generate a decant stream and settled-solids stream. The decant stream may be recycled to the plant influent or potentially discharged to a local water body.

Discharging the decant stream to a local water body depends on the local regulatory conditions. Recycling the decant to plant influent is common practice for iron and manganese treatment plants in the mid-Atlantic region of the United States. The solids stream generated from the clarification of spent filter backwash will be 10 percent to 15 percent of the total spent filter backwash volume. The solids stream can be hauled off site or dewatered.

There are two primary methods of clarifying spent filter backwash residuals: (1) batch settling and decant and (2) equalization and continuous treatment.

Batch settling and decant consist of a tank that contains one or more backwash volumes where the backwash waste settles, followed by decanting. Generally the spent-filter backwash from an iron and manganese pressure filter will settle in approximately 3 hr, after which the decant quality is generally less than 1 ntu. After the settling period is completed, decant water is pumped to the plant influent or stream. The sludge is allowed to accumulate in the bottom of the tank and after several backwashes is pumped either to the dewatering system or to a tanker truck. Figure 8-2 presents a schematic of typical batch-settling tank.

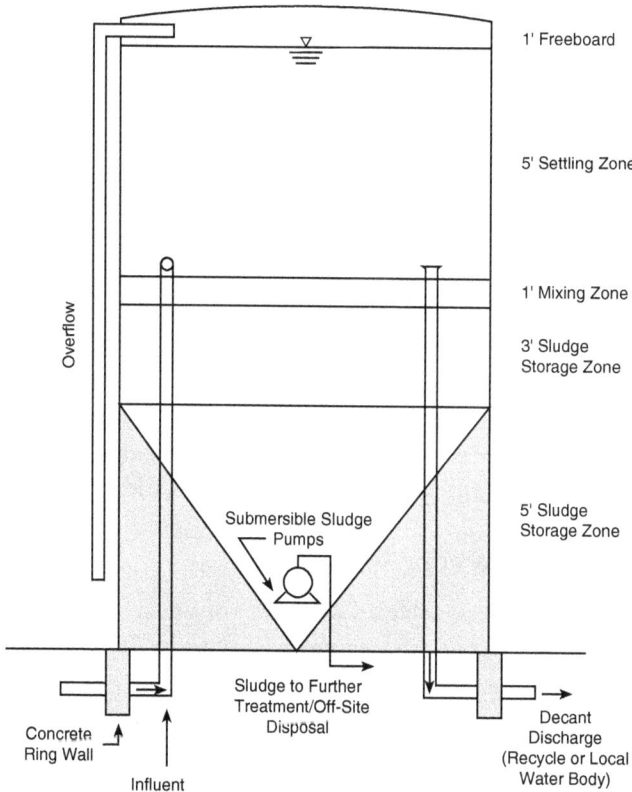

Figure 8-2   Typical batch-settling tank
Source: Hatch Mott MacDonald

The number of batch tanks depends on the number of filters and frequency of backwash. For smaller-capacity facilities that have redundant sources of supply, one tank may be sufficient. For larger facilities, a second tank will generally be needed. Note that treatment plants with a capacity of 10 mgd and with 10 or more pressure filters generally need only two tanks, but they must be larger in capacity.

Equalization and continuous treatment. An advantage of this method is that the decant stream is generated continuously, which reduces the impact on the treatment plant when the decant stream is recycled. This system consists of an equalization tank generally sized for two backwash volumes, along with a clarification process. Plate settlers are generally effective because their footprint is much smaller than that of a conventional clarifier.

Providing treatment to allow recycling of the decant streams significantly reduces operational costs compared with disposing of the residuals to the sewer. The only operating costs associated with the treatment systems is the energy associated with operating the pumping and collector systems. Because the pressures are very low, the energy costs are typically very low. However, recycling of decant streams results in a change in the feedwater to the plant and relies on the clarification system at the influent of the plant and the spent filter backwash clarification system to remove solids. During periods of upsets to the clarification system (perhaps caused by a change in raw water chemistry or an algal episode), the clarifiers may not be able to sufficiently remove these substances, which may be continuously recycled, resulting in short filter runs. The capability to discharge the decant to a sewer or a local water body provides additional operational flexibility.

Nonmechanical Dewatering

If sewer disposal is not possible and trucking of thickened solids off site is not cost-effective, the residuals will need dewatering. The term *dewatering* generally refers to the process of producing cake that has no free liquid. The paint filter test is used determine whether solids can be disposed of in a landfill and is determined from Method 9095B, the Paint Filter Liquids Test, which states, "A predetermined amount of material is placed in a paint filter. If any portion of the material passes through and drops from the filter within the 5-min test period, the material is deemed to contain free liquids" (USEPA 2004).

Generally the minimum solids concentration is 15 to 18 percent. Cake disposal costs are generally based on a dollar per wet ton basis. Therefore, the dryer the cake, the lower the cost of disposal.

Nonmechanical methods include sand drying beds and lagoons. The feasibility of nonmechanical methods depends on two primary factors: climate and available land. Facilities with acres of space can potentially accommodate drying beds. Climates that receive significant amounts of rainfall are less suitable for drying beds, but the beds can be enhanced by providing decanting, underdrain systems, and roofs. Drying beds entail the application of liquid wastes to a surface area that typically contains sand-type filter media. Filtered water can be collected and recycled or it can be discharged to the groundwater. After several months the residuals will be dewatered and can be removed using an excavator. Nonmechanical dewatering is capable of achieving cake solids as high as 70 percent.

Mechanical Dewatering

In many instances, space is not available for the construction of non-mechanical dewatering. Mechanical dewatering consist of using a machine such as a belt press, plate and frame press, centrifuge, volute press, or screw press to produce a cake. The cake produced from a mechanical dewatering system treating iron and manganese residuals will generally be 15 percent to 30 percent cake solids. Most of the treatment technologies require the application of a polymer to generate the cake. Figure 8-3 illustrates a centrifuge used for dewatering iron and manganese residuals.

To select the most appropriate mechanical dewatering system, bench testing various technologies is recommended to assess the effectiveness of each. Bench testing should be followed by pilot testing of the most viable treatment technology to assess the optimum loading rates for the system.

Mechanical dewatering systems generally use a polymer to enable dewatering of the residuals, and the mechanical dewatering system can generate a liquid waste stream that contains residual polymer. Direct recycle of the mechanical dewatering liquid stream should be avoided because the polymer could adversely impact the main treatment process. The liquid stream can be conveyed to the thickener or spent filter backwash treatment system.

Design Example

For the 700-gpm well design example with 2 mg/L of iron and 0.5 mg/L of manganese used in the chapter 7 design example for manganese dioxide–coated sand, the conceptual design of the filtration system consisted of the following:

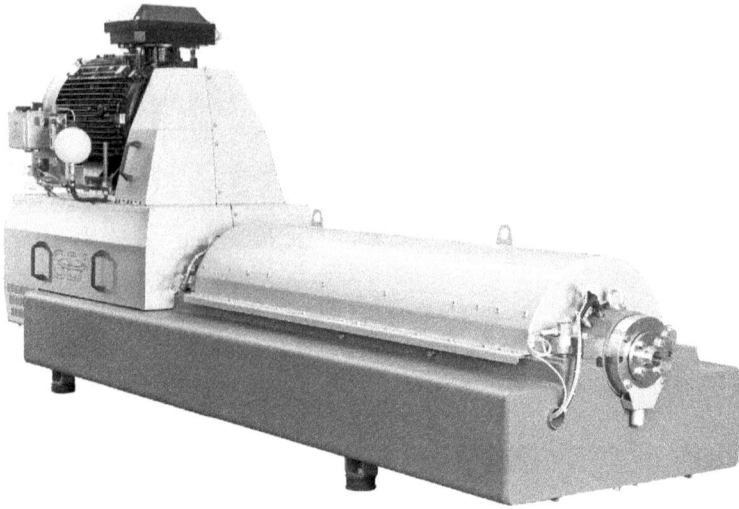

Figure 8-3   Centrifuge for dewatering iron and manganese residuals
Courtesy of GEA Mechanical Equipment

Type of operation: catalytic oxidation

Number of filters: 3, each 350-gpm capacity
(2 duty/1 standby)

Design loading rate maximum: 6 gpm/ft^2

Diameter per filter: 9 ft

Area per filter: 63 ft^2

Depth of anthracite: 18 in.

Depth of manganese dioxide–coated sand media: 18 in.

Iron and manganese concentration: 0.17 grains/gal

Capacity of manganese dioxide–coated sand: 900 grains/ft^2

Treated volume per filter: 333,529 gal

Run time per filter: 16 hr

Backwash volume: 12 gpm/ft^2 × 63 ft^2 × 10 min = 7,630 gal

Residuals computations for sewer discharge, sewer and equalization, and batch settling are presented here.

Sewer discharge. The instantaneous backwash rate is 12 gpm/ft^2 × 63.6 ft^2 = 763 gpm. For direct sewer discharge to be feasible, the capacity of the sanitary sewer system needs to be a minimum of 763 gpm plus capacity for any other discharge flows from the facility.

Sewer with equalization. The total backwash from the two duty filters is 7,630 gal/filter × 2 filters = 15,260 gal. With a 16-hr filter run, the daily volume of spent filter backwash would be 15,260 × 24 ÷ 16 = 23,000 gal. The continuous equalized flow rate to the sewer could be limited to approximately 16 gpm (23,000 gpd ÷ (24 hrs/day × 60 min/hr). The equalization tank volume would need to be a minimum of 7,630 gal. There would be approximately 8 hr between backwashes.

Batch settling and decant. The volume of settled sludge generated by batch settling would be approximately 15 percent of the total filter backwash volume sent to the holding tank, and the remaining 85 percent would be clarified water that could be decanted. The backwash settling tanks could be designed to have a settling volume for two filter backwashes along with three days of sludge storage.

For this scenario, the volume of settled sludge from two backwashes would be approximately 0.15 × 15,260 gal = 2,290 gal, and the volume of water from the two backwashes that could be decanted would be approximately 0.85 × 15,260 = 12,971 gal. The volume required for three days of sludge storage for the two duty filters would be approximately 7,000 gal, and the volume required for two backwashes would be approximately 16,000 gal (7,630 gallons per backwash × 2 backwashes = 15,260 gallons).

The minimum size of the backwash tank for this scenario would therefore be approximately 23,000 gal (2 backwashes × 7,630 gal/backwash + 3 days sludge × 2,290 gal of sludge per day). If filter backwashes were performed back to back, the backwash waste could be decanted at a rate of 30 gpm. This would allow for decanting to be completed in approximately 7 hr. If the decant were recycled back to the head of the treatment process, this decant flow rate would represent less than 5 percent of the overall plant flow. The settled sludge in the tank would need to be pumped to further treatment or a truck for hauling a minimum of every three days.

SUSTAINABILITY

Before sustainability became a performance metric, the practice for residuals handling was to implement the most expedient option regardless of long-term consequences. These options have included direct discharge to watercourses, storage of residuals on site in unlined lagoons, and landfilling of wastes.

In highly developed areas of the east and west coasts of the United States, less landfill space is available, requiring the residuals to be hauled much longer distances at greater cost. Sewer systems are confronted with similar issues regarding the limited options available for residuals disposal. While the iron and manganese residuals do not adversely impact most wastewater treatment processes, these inert materials make the wastewater treatment plant residuals less amenable to uses such as fertilizer supplements and energy production.

Improving the sustainability of the residuals system embodies the traditional 3R approach: Reduce, Reuse, Recycle. Reduction of residuals production through optimization of the source water and clarification and filtration systems should be considered. If dewatering of the residuals is provided, increasing the solids concentration of the cake will provide a similar benefit. Recycling of the residuals, as discussed earlier, improves the overall efficiency and sustainability of the treatment facility, but caution should be exercised when water quality changes or there are treatment process upsets. The capacity or the finished-water quality should never be compromised to achieve sustainability goals.

To reduce the environmental impact of residuals, future treatment technologies in conjunction with management practice changes may provide alternate uses of residuals generated from iron and manganese facilities. In the short term, operational adjustments can be made to reduce the amount of residuals generated. Residuals minimization techniques include optimizing the backwash systems to reduce the volume of residuals generated and treatment of the spent filter backwash with recycling of the decant stream.

9

Distribution System Water Quality

A significant number of customer complaints are prompted by dirty water resulting from the internal corrosion of unlined cast-iron water mains. The composition of the "dirty" water is most often associated with a release of iron corrosion by-products. This chapter presents an overview of distribution system corrosion related to iron, water quality factors that increase iron release, physical factors that can increase iron release, and a summary of operational and rehabilitation/replacement techniques to reduce discolored water.

IMPORTANCE OF DISCOLORED WATER TO OPERATIONS

Discolored water from a customer's tap is an aesthetic issue but not necessarily a regulatory violation. However, the presence of discolored water from tap samples can lead to decreased consumer confidence and adversely impact the public's opinion of a water utility. Discolored water can also be symptomatic of deeper water quality problems, such as low disinfectant residuals; elevated tuberculation, which could reduce the carrying capacity of the piping; and internal piping corrosion, which could negatively impact the structural integrity of the piping. Figures 9-1a and 9-1b illustrate unlined and lined cast-iron piping from the 1940s.

Water utilities should consider discolored water complaints as part of the larger goal of optimizing distribution system performance, incorporating the complaint reports into metrics such as maintaining minimum pressure, reducing main breaks, and providing adequate fire flows. An integrated approach can allow for all of the issues to be addressed using a combination of capital and operational improvements. For example, discolored water may be occurring in areas with older, unlined cast-iron pipe, some sections of which have experienced significant main breaks. In those areas with main breaks, pipes may be replaced or a structural lining may

Figure 9-1a   Unlined cast-iron piping
Source: Hatch Mott MacDonald

Figure 9-1b   Lined cast-iron piping
Source: Hatch Mott MacDonald

be applied. In the areas that have not had main breaks, flushing could be performed, or the pipes could be rehabilitated by cleaning and lining them, eliminating the sources of dirty water and the related complaints.

When iron is released from corrosion scales in iron pipe, it goes into suspension and gives drinking water a red, brown, or yellow color or a dirty appearance. The chemistry of corrosion scale reactions that lead to discolored water is complex, and research is still being conducted. This section provides a summary of relevant current scientific knowledge about the processes underlying discolored-water occurrences and how water quality, hydraulic conditions, and microbial activity affect the water chemistry.

The processes of iron corrosion, pipe scale development, and iron release are highly interrelated. A reasonable understanding of how water quality, hydraulic conditions, and microbial activity affect the overall reaction system is needed to properly identify the cause(s) for discolored water problems so that management can make informed decisions that will minimize these problems sustainably, without causing unintended consequences for other important drinking water goals. Utility management and operational staff need to understand the balance between water quality and operational adjustments before changes are implemented to address discolored-water quality issues caused by release of iron from corroded pipes.

First, the culprit pipe should be identified. The occurrence of discolored water at a specific location may be separate in time and location from related corrosion and scale development processes. Water quality conditions at the time of dirty-water occurrences may not be directly related to iron release rates, and study results may be ambiguous if distinctions are not made between the actual corrosion and iron release into the tap water. Monitoring studies and data collection are discussed later in this chapter.

SCALE FORMATION AND STRUCTURE

Most iron-corrosion scale originates in a corroded layer on the pipe wall. The corrosion by-product (tuberculation) has a porous interior covered by a relatively dense, hard, somewhat impermeable shell-like layer that provides limited structural integrity and a loosely attached top surface layer (Sarin et al. 2004a). Figure 9-2 presents a schematic diagram of the types of scales that can form in an unlined cast-iron pipe.

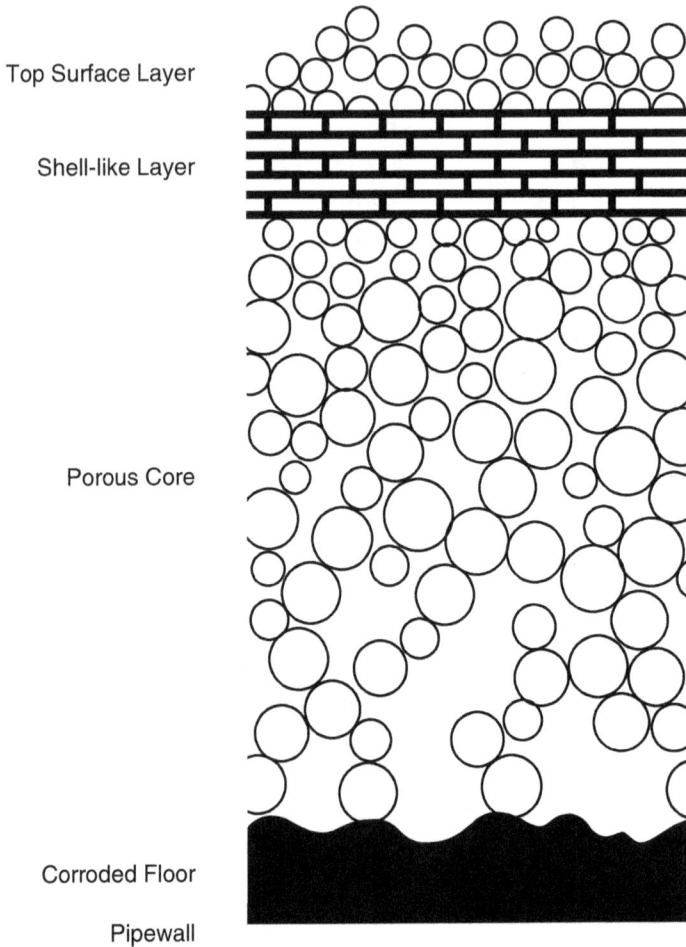

Top Surface Layer

Shell-like Layer

Porous Core

Corroded Floor

Pipewall

Figure 9-2   Schematic of scale formation in unlined cast-iron piping

Cavities in the scale interior (core) may result from acidic conditions that prevent iron oxide and hydroxide precipitation. The scale core generally contains masses of small particles of variable shapes, with the minerals of goethite, magnetite, maghemite, and siderite having been identified.

Multiple shell-like layers may form due to successive fractures caused by thermal expansion and contraction, or from the force of corrosion product buildup in the scale interior. Dense thick shell-like layers result from slow tubercle growth, whereas thinner layers result from faster growth. Dense shell-like layers retard corrosion by providing a barrier to inward diffusion of oxygen and charge-balancing anions that support continued corrosion. Surface layers covering the shell-like layers interface with the bulk water and are thus strongly influenced by water quality conditions.

Initial corrosion rates on clean or new iron pipe decrease rapidly as scale grows and provides increased resistance to corrosion reactions. Iron scale formation is complex since the corroded metal can exist in several oxidation states (II, III, and mixed II/III states) and in more than a dozen different soluble or precipitated forms (AwwaRF and DVGW-TZW 1996). The composition and structure of iron scales depend on the metal type, water constituents (e.g., pH, alkalinity, buffer intensity, natural organic matter [NOM], dissolved oxygen, and corrosion inhibitors), hydraulic flow patterns, temperature, and microbiological activity.

Hydraulic surges and turbulent flow (e.g., during high demands such as fire fighting or main breaks) can dislodge parts of the loosely attached outer layers into the bulk water, causing turbidity and discolored water. When oxidant concentrations at the scale surface are low or depleted, Fe^{2+} ions that reach the interface can diffuse into the bulk water, where they will subsequently precipitate as Fe(III) particles when an oxidant is encountered, contributing to discolored water.

WATER QUALITY FACTORS

Corrosion rates, scale solubility, and oxidation rates are all influenced by water quality conditions. Changes in water quality impact the development of scale mineral properties. Water quality parameters affecting corrosion include oxidants, alkalinity, pH, chlorides, sulfate, and microbial activity.

Oxidants and ORP Variability

Dissolved oxygen (DO) is consumed in the corrosion reaction and by Fe^{2+} oxidation. High DO concentrations can lower corrosion and iron-release rates by oxidizing soluble scale components, which leads to denser and less porous scale that better impedes oxygen and ion transport. Higher flow rates increase the diffusion of oxidants from the bulk water to the scale surface where oxidation can intercept Fe^{2+} and thus decrease iron release.

Higher DO concentrations enhance initial stages of corrosion (before scales are developed) and may increase corrosion rates on old pipes (and possibly subsequent iron release) if scales are highly porous and conductive. However, corrosion can still proceed under anoxic conditions.

Alkalinity and pH

Higher alkalinity and/or buffer capacity may reduce corrosion rates by stabilizing local pH conditions at corrosion sites. Stabilizing the pH may also reduce the development of corrosion concentration cells. Denser scales are associated with higher alkalinity waters, consistent with this model (Sarin et al. 2001).

Scales from a low-alkalinity water supply (10 to 35 mg/L as $CaCO_3$) in Boston were found to be more porous and to have larger pores than scales from a high-alkalinity midwestern system (135 to 150 mg/L as $CaCO_3$). An increase of pH from 7.5 to pH 9.5 in a distribution system with old corroded cast-iron pipes led to a gradual reduction of iron release, indicating that the effect was attributable to beneficial transformations in scale microstructure (Sarin et al. 2003).

Chloride, Sulfate, and Conductivity

Chloride and sulfate anions facilitate corrosion through their role as charge-balancing anions and by fostering acidic conditions that promote primary corrosion cell reactions (as well as corrosion scale dissolution).

The chloride anion is particularly mobile through scale due to its small ionic radius. In contrast to chloride and sulfate, bicarbonate anions can also provide charge balance for corrosion reactions. At the same time, bicarbonate anions are buffering, so they neutralize corrosion-generated H^+ and OH^- ions with the beneficial results discussed in the previous section. Therefore a higher ratio of chloride and sulfate to bicarbonate anions is expected to enhance corrosion, increase scale porosity, and lead to higher iron-release rates.

Phosphate

The mechanism of action of how phosphate inhibitors work with respect to iron corrosion and release is an area of ongoing research. Phosphate products are usually employed for lead and copper corrosion control, and most of the research addresses these applications.

Orthophosphates are believed to control iron release in corroded cast-iron pipes by precipitating iron phosphates that fill pores in the top surface or shell-like scale layers, thereby reducing scale porosity (Sarin et al. 2003). Calcium carbonate precipitates are thought to play a similar role. Orthophosphate addition partially ameliorated the rise in iron and turbidity caused by chloride addition in experiments using old cast-iron pipes under stagnant water conditions (Lytle et al. 2003). Clement et al. (2002) found that orthophosphates at a residual of 3 mg/L reduced iron release from old unlined cast-iron pipes in low-alkalinity waters (30–35 mg/L as $CaCO_3$) but had little effect in high-alkalinity waters.

Questions are often raised about whether phosphate addition for corrosion control might enhance bacterial growth in the distribution system by supplying a limited nutrient. Study findings indicate this to be untrue in most cases. One laboratory study reaching this conclusion reported an immediate and drastic drop in both iron release and bacterial production after phosphate addition to a highly corroded unlined cast-iron reactor and a large concomitant uptake of phosphate by the pipe (Appenzeller et al. 2001).

Temperature

Higher water temperatures increase microbiological activity, chemical reaction rates, and diffusion transport rates, and decrease oxygen solubility. These effects would be expected to have mostly negative consequences for discolored water. The lack of oxidants at the scale surface has a direct effect on release of Fe^{2+} into the bulk water, as described previously.

Pipe metal and attached scales also undergo thermal expansion and contraction from temperature changes, which may cause hard scales to crack (Sarin et al. 2004). Studies have shown that iron release may increase with temperature, but the overall effect is uncertain and may be system-specific (McNeill and Edwards 2001, 2002). Larson (1975) noted that higher temperature, will increase the availability of all water constituents at the pipe wall, whether corrosive or protective. The net effect of temperature on corrosion reaction processes depends on which processes are dominant, according to the water composition.

Microbial Impacts

A wide array of microorganisms may colonize favorable environments in the distribution system. Oxygen concentration cells induced by microbiological activity are an important cause of corrosion in distribution networks, and microbial activity consumes oxygen and chlorine residual. Known examples include iron oxidation, iron reduction, sulfur oxidation, sulfate reduction, ammonia oxidation, and nitrite oxidation. In general, the reduction processes proceed in anoxic environments and tend to raise pH, whereas the oxidation processes consume DO and tend to lower pH.

The lack of a shell-like layer in some scales has been attributed to iron-oxidizing bacteria causing porous scale structure (Sarin et al. 2004b).

Corroded-pipe scales harbor biofilms, which will entrain iron corrosion by-products whether or not the corrosion is microbiologically mediated. Biofilms will transport these products into the bulk water when they slough off pipes.

Factors Influencing Discolored Water Processes

Discolored water from customer taps is the result of several physical and chemical interactions. Figure 9-3 illustrates the complex interactions associated with iron corrosion and discolored water.

Hydraulics is associated with the velocity of water moving through the piping network along with water age. Low velocities can allow particles to settle and accumulate and then become resuspended during changes in demand patterns. Flushing programs can be used to remove the lighter particles that have accumulated, and the velocity will depend on the physical nature of these particles. Hydraulics also impact water quality parameters, such as chlorine residual, by affecting the residence time (age) of the water in the piping network.

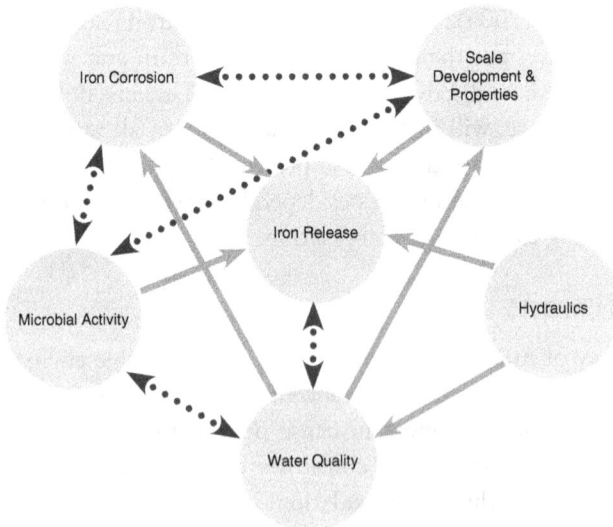

Figure 9-3   Interacting processes and impacts. Solid arrow = one-way contribution. Dotted arrow = two-way contribution.

Water quality can impact discolored water and iron release. Parameters such as conductivity, pH, alkalinity, and hardness can change seasonally as well as vary throughout the distribution system. Understanding which parameters have the most significant impact on discolored water is site-specific, and the next section provides an overview of the relevant water quality parameters that should be considered.

Microbial activity is directly impacted by water quality parameters and can also lead to iron release. Microbial activity is generally more pronounced in chloraminated systems where microbial-induced corrosion can occur. Microbial activity can be mitigated by managing water quality at the point of entry and by minimizing water age.

Iron corrosion of the pipe wall is influenced by water quality parameters as well as the type and extent of microbial activity. However, corrosion of the pipe wall will not necessarily result in discolored water. Corrosion of the pipe wall will lead to the formation of various types of scales, each with its own unique chemical and physical properties.

Scale development and properties will vary depending on location within the distribution system and will change over time due to variations in water quality and hydraulic conditions. It is important to understand which types of scale result in the release of iron, along with the rate of scale development and the factors that influence scale development.

IMPLEMENTING SOLUTIONS TO ADDRESS DISCOLORED WATER

To address discolored-water issues, a robust monitoring program should be in place along with operational and capital improvements. Simply put, understanding the water quality parameters with respect to time and space and then implementing operational and physical improvements can reduce discolored water.

Water Quality Data

The first step in addressing the issue of discolored water in the distribution system is to have data on distribution system water quality and the condition of the piping network. The data should be collected and organized to simplify analyses. Sample sites should be selected from problematic areas as well as areas that do not exhibit discolored water, to provide a basis for comparison. Analysis of the data using GIS is often helpful in identifying spatial and temporal correlations.

Sampling should be performed on a regular basis. For systems in which temperature and demand can vary seasonally, samples should be collected every quarter.

The following water quality parameters should be considered for analyses:

- Alkalinity
- Aluminum
- Ammonia: free (for chloraminated systems)
- Ammonia: total (for chloraminated systems)
- Chloride ion
- Calcium hardness
- Total hardness
- True and apparent color
- Conductivity
- Chlorine: free
- Chlorine: total
- Monochloramine (in chloraminated systems)
- Heterotrophic plate count (HPC)
- Iron
- Nitrite (in chloraminated systems)
- Orthophosphate
- Orthophosphate: dissolved
- Oxidation reduction potential (ORP)
- pH
- Sulfate
- Temperature
- Total dissolved solids (TDS)

Operational Improvements

Operational improvements to reduce discolored water include conventional flushing, unidirectional flushing, swabbing, and innovative technologies such as neutral output discharge elimination and ice pigging. Whenever operational improvements are being implemented, water quality should be monitored to assess the effectiveness of the improvements.

Conventional intermittent flushing can be performed to address short-term and site-specific issues, but it is generally not a long-term solution for systems with large amounts of unlined cast-iron piping.

Unidirectional flushing (UDF) involves flushing using a specific sequence of valve closures that allow for higher flushing velocities to scour the water main. UDF pulls water through one segment of main at a time. The piping is cleaned progressively, starting with the treatment plant and moving outward to the ends of the system. Valves are closed strategically to direct the flow of water from clean pipes into dirty ones. The UDF program involves a significant amount of water quality monitoring to assess the effectiveness of the flushing. The required frequency of UDF should be determined on a case-by-case basis.

Swabbing involves passing a foam plug (pig) through the distribution lines. Workers insert a compressible plug larger in diameter than the lines to be cleaned into a hydrant or other point designed for that purpose. The pig is forced down the insertion line or hydrant using pressure from a fire department pumper truck or a portable pressure pump. Distribution line pressure then carries the plug to a hydrant that has had its internal mechanism removed so the plug can exit.

The Neutral Output Discharge Elimination System (NO-DES) flushing method is an environmentally friendly alternative to conventional water main flushing (NO-DES 2015). Water quality is improved without wasting millions of gallons of potable water. Instead of water flowing out of fire hydrants and running to waste, the NO-DES process uses a trailer-mounted pumping, filtering, and rechlorinating system that circulates the water within the water distribution system. The NO-DES unit is connected with large hoses between two fire hydrants, creating a temporary loop in the water distribution system. A pump on the unit circulates water at prescribed velocities, passing it through a series of filters that remove intentionally stirred-up sediment and particulate matter. Disinfectant can be added to further improve water quality and ensure its safety. Inline turbidity meters indicate when desired clarity levels are met, providing site-specific data to determine when the flushing has achieved its purpose. Figure 9-4 presents a schematic of the NO-DES process.

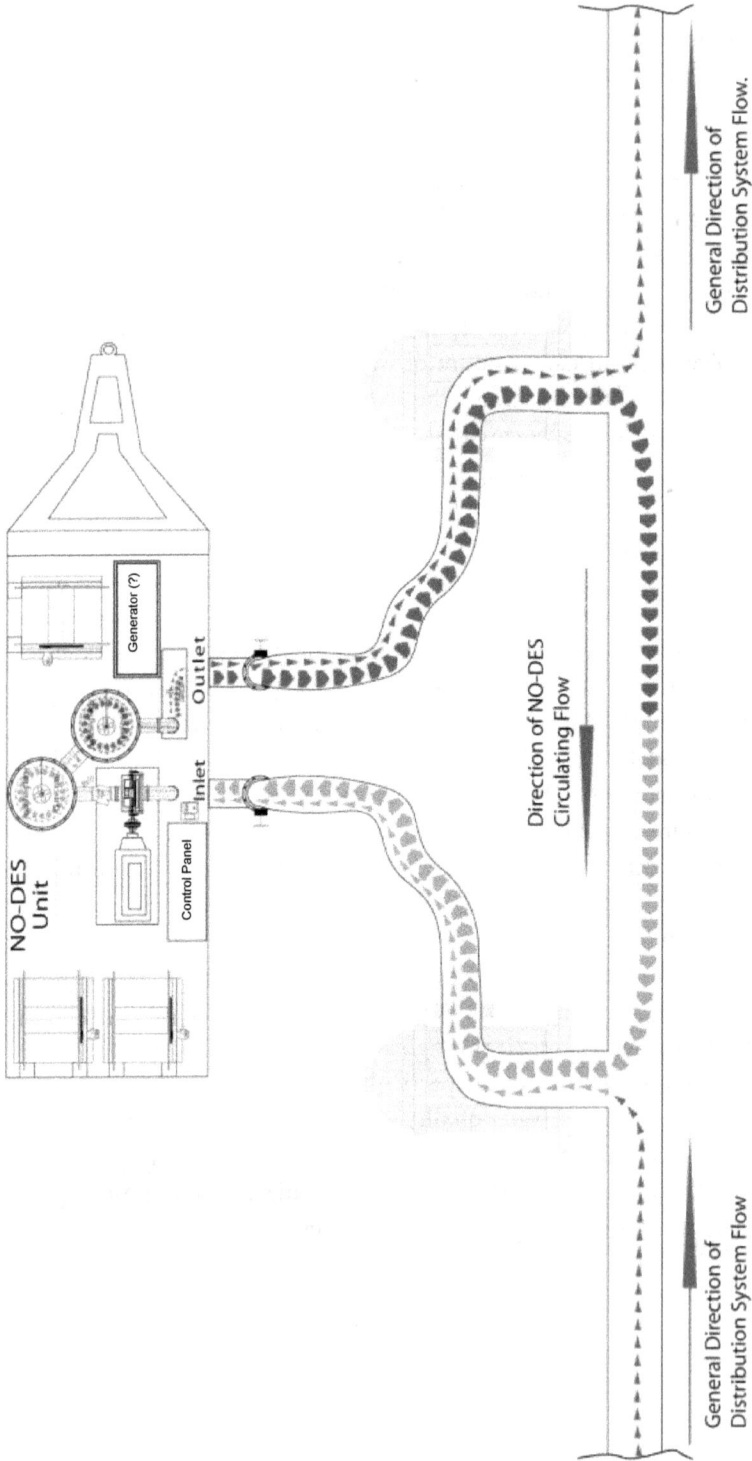

NO-DES Unit

Generator (?)

Control Panel

Outlet

Inlet

General Direction of Distribution System Flow

Direction of NO-DES Circulating Flow

General Direction of Distribution System Flow

Figure 9-4   Schematic of NO-DES process
Courtesy of NO-DES Inc.

Advantages of the NO-DES method include the following:

- Elimination of most of the water waste associated with conventional flushing
- Removal of particulates, biofilm, iron, and manganese
- Elimination of NPDES-related issues
- No property damage or cleanup costs; no auto accident claims
- Allows flushing of mains after main repairs to avoid contamination
- Water quality improvements
- Regular flushing, 24/7/365
- Controlled flow and reduced damage to distribution system—no water hammer, pipe/lining erosion
- Lowered labor costs because of the smaller crew, reduced setup and valving
- Dechlorination not required
- Energy savings—pumping costs for lost water, pipe capacity regained
- Consistent message to public—conservation, no misperceptions about water quality due to observance of dirty water

Ice pigging combines the operational advantages of flushing with the cleaning impact of soft pigging. The ice pig is a semisolid that is pumped like a liquid and flows without blockage through changes in diameter, bends, and fittings. Ice pigging has a minimum impact on operations because it is simply pumped into and recovered from a hydrant at each end of the pipe section without excavation or modification to the hydrant (Utility Service Group 2014).

Whenever any of these operational strategies are employed, care should be taken to avoid disruption to the distribution system. If the utility has no experience with the procedure, a pilot program should be established to gain experience and assess the effectiveness of the technology.

In some cases, operational improvements will provide only temporary relief of discoloration. All of these operational improvements require labor to perform the activities. Even if the work is performed by a contractor there will be costs associated with the program. Flushing programs are also not permanent fixes; for a permanent solution to discolored-water problems, capital improvements may be needed.

Capital Improvements

Industry best practice indicates that the estimated life of buried cast-iron piping is approximately 100 years. However, this life could be significantly shortened depending on the environmental conditions the pipe is subjected to—e.g., corrosive soils, stray currents, corrosive water, and other factors. Many systems with older unlined piping are faced with addressing short-term water quality issues caused in part by the unlined piping, while needing to optimize the remaining useful life of the piping.

Replacement of the unlined piping provides the most robust solution, but is also the most expensive option. A less expensive option is lining the existing piping with nonstructural, semistructural, and structural liners.

Nonstructural liners were traditionally limited to cement-mortar lining. The cement lining would not only improve water quality by eliminating the tuberculation but would also restore the full carrying capacity of the piping. However, the cement-mortar lining does not increase the structural strength of the piping. Semistructural and structural liners may be less expensive than pipe replacement and can allow a utility to rehabilitate a larger portion of its distribution system for the same price.

Semistructural liners include spray-on polymer/sprayed-in-place pipe (SIPP) systems. SIPP systems are available in different chemical compositions and include polyurethane and polyurea-based polymer linings and hybrids formed by the combination of the two. These materials have a rapid cure time and the ability to structurally enhance the host pipe. Some products bond to the host pipe while others offer independent structural support. This type of product is cost-competitive when compared with cement-mortar products, and offers the advantage of potentially extending the service of the pipe.

Fully structural liners are generally of the cured-in-place pipe (CIPP) type. The CIPP lining technique consists of a resin-saturated nonwoven polyester felt or woven polyester fiber tube impregnated and/or coated with a thermoset resin. Once the tube is inserted into the pipe, it is cured under ambient conditions, using hot water or steam to form a pipe within a pipe.

SUSTAINABILITY

The piping in most water distribution systems varies in terms of pipe material (unlined cast iron, lined ductile iron, PVC, HDPE, etc). The internal condition of these pipes will vary based on the date they were installed, the impact of water quality on the development of scales, and other factors.

Sustainable solutions for reducing discolored water involve implementing options that address immediate needs while viewing the system through a long-term prism. Reducing the corrosivity of the water caused by the corrosion of water mains is the first part of a sustainable management strategy, as it eliminates energy and water usage associated with flushing and the disruption to the system associated with pipe replacement or the use of liners.

Flushing is often needed to address immediate discolored-water issues. However, flushing requires use of nonrevenue water as well as the use of energy and chemicals for the production of treated water that is not available to the customer. New flushing techniques such as NO-DES address some of the sustainability issues associated with conventional flushing.

Pipe replacement provides a long-term solution in areas where the pipes are heavily tuberculated and may have structural problems. However, replacement results in more intense construction activities that generate a larger climate footprint than lining technologies. While lining technologies will generally involve less surface construction than pipe replacement, many of the liners do not have the longevity of conventional ductile-iron pipe. The use of a sustainability rating system described in chapter 2 may be beneficial as part of the assessment of options to address discolored water.

10

Case Studies

Case studies are useful because they apply the information from the previous chapters to actual projects that deliver water that is low in iron and manganese. Each case study includes the following:

- An introduction to the overall project
- A summary of the raw water quality conditions
- Finished water quality goals
- Details on the iron and manganese removal process

As discussed in this book, the treatment of iron and manganese can be achieved using different methods based on various water quality parameters and concentrations of iron and manganese. The design examples presented in this chapter demonstrate different treatment system designs. Table 10-1 provides a summary of the design basis for the examples.

CASE STUDY NO. 1: SPRINGFIELD WELL FIELD

Introduction

New Jersey American Water (NJAW) owns and operates a well supply and pumping system known as the Springfield Well Field. The Springfield system dates back to 1907 and originally consisted of up to 50 wells that pumped to a centralized facility for chlorination, followed by booster pumping to the distribution system. Over many years, wells were removed from the system due to a variety of problems, and in 1989 NJAW discontinued the use of all the remaining wells because of water quality concerns. Based on the availability of water supply and the growing needs of the system, NJAW decided to reactivate the well supply in the early 1990s.

Table 10-1   Design and selection summary for case studies

Example Project	Design Process	Reason for Process Selection
Case Study No. 1	Split train of weak acid cation exchange (WAC) and manganese greensand (catalytic oxidation), both using vertical filters followed by packed tower aeration (PTA)	Hardness reduction (WAC) and manganese removal CO_2 removal (PTA)
Case Study No. 2	Manganese greensand pressure filtration using horizontal dual cell filters	Iron-related bacteria (IRB) (*Crenothrix*) removal Low-level iron and high-level manganese removal
Case Study No. 3	Aeration, pH adjustment, solids-contact clarification, dual media (sand/anthracite) pressure filtration	High levels of iron
Case Study No. 4	Manganese greensand pressure filtration using vertical filters	Low levels of iron, moderate levels of manganese

Treatability studies conducted in the 1990s indicated that various forms of treatment would be required to return the well field to operation. Studies recommended an overall treatment process of weak acid cation (WAC) exchange for hardness reduction, manganese greensand pressure filters for iron and manganese removal, packed tower aeration (PTA) for the removal of CO_2, and final disinfection with sodium hypochlorite. Although several treatment techniques were required to adequately meet all the finished water quality goals, only the treatment for iron and manganese removal is discussed in this case study. The remaining systems will be discussed only as necessary to provide a complete overview of the treatment system. As part of this project, 13 of the remote well sites were reactivated. The overall system capacity is 4 mgd.

Water Quality

Although each well has a unique water quality, the wells are operated concurrently and discharge into a common raw water wetwell. The composite water quality used for the design of the treatment plant is summarized in Table 10-2.

The finished water quality goals for this treatment plant were determined to meet New Jersey Department of Environmental Protection standards. Hardness and alkalinity goals were established by NJAW based on operational requirements within the distribution system. A summary of the finished water quality goals is depicted in Table 10-3.

Treatment Technique

The treatment process at the Springfield Water Treatment Plant (WTP) consists of a split-stream system. In order to meet the finished water quality goals for total hardness, only a portion of the raw water needs to be treated. Approximately 28 percent (1.12 mgd) of the raw water flow is treated with the WAC system, which also removes the iron and manganese from this portion of the flow. The remaining 72 percent (2.88 mgd) is treated with the manganese greensand pressure-filtration system. The entire flow is then recombined and treated in the packed tower aeration system for the removal of CO_2. The manganese greensand system for this WTP was designed around equipment provided by Hungerford & Terry (H&T).

Table 10-2 Composite raw water quality for NJAW well supply

Parameter	Composite Raw Water Concentration
Iron	0.16 mg/L
Manganese	0.11 mg/L
Total hardness	235 mg/L as $CaCO_3$
Alkalinity	123 mg/L as $CaCO_3$
Total dissolved solids	344 mg/L

Table 10-3 Finished water quality goals for Springfield Water Treatment Plant

Parameter	Finished Water Quality
Iron	<0.16 mg/L
Manganese	0.05 mg/L
Total hardness	200 mg/L as $CaCO_3$
Alkalinity	80 mg/L as $CaCO_3$
Total dissolved solids	<344 mg/L

Prior to installation of the system, H&T was contacted and provided with the raw water quality and treated water quality goals so tests could be run to determine the best equipment for the job. Manganese greensand systems can operate in three distinct modes, each of which was investigated to optimize the system operation. The first mode of operation evaluated was continuous regeneration (CR). CR is recommended for water where iron removal is the main objective. This mode of operation requires a continuous feed of potassium permanganate and chlorine prior to the filters. The chlorine oxidizes the iron while the potassium permanganate oxidizes the manganese and regenerates the greensand media. This mode typically operates with a liquid loading rate of approximately 5 gpm/ft^2.

The second mode of operation evaluated was intermittent regeneration (IR). IR is typically used where only manganese or manganese with small amounts of iron are in the water. This method involves feeding potassium permanganate only when the media bed needs regeneration. The liquid loading rate for this mode of operation is also approximately 5 gpm/ft^2. Since both CR and IR modes contain potassium permanganate in the backwash wastewater, the local sewer authority was consulted during the evaluation because it would be receiving the waste.

The third mode of operation evaluated was catalytic oxidation (CO). CO is typically used with relatively low levels of iron and manganese in the water. The CO mode does not require feeding any potassium permanganate once the system has been commissioned. During startup, a small amount of potassium permanganate (approximately 5 gal per vessel) is used to coat the media. A chlorine solution is added to the raw water prior to the filters at a level that results in a measurable free chlorine residual after the filter. The liquid loading rate for CO mode can be as high as 8 gpm/ft^2 or greater.

Given the low levels of both iron and manganese in the raw water, the catalytic oxidation mode of operation was selected. Table 10-4 presents a summary of treatment system characteristics.

As with any filtration system, each filter must periodically be backwashed to remove particulate matter. The greensand system at Springfield is backwashed at a rate of 12 gpm/ft^2. Typical backwashing rates are 12–15 gpm/ft^2. This flow rate provides for sufficient bed expansion to achieve good cleaning. Every fifth backwash cycle is augmented with air scouring to enhance cleaning of the media.

Table 10-4 Greensand system characteristics summary

Type of Vessel	Vertical, Single Cell
Number of vessels	3
Vessel diameter (ft)	11
Surface area per vessel (ft^2)	95
Maximum flow rate per vessel (gpm)	667
Maximum liquid loading rate (gpm/ft^2)	7.0
Volume of greensand per vessel (ft^3)	142
Depth of greensand media (in.)	18
Depth of anthracite media (in.)	18
Maximum head loss across system (psi)	10

A backwash cycle consists of several distinct steps. The first step is a drain-down to waste to provide space within the vessel for bed expansion. The second step consists of the backwash sequence at 12 gpm/ft^2. The third step is a rinse to waste. The backwash cycle is initiated automatically based on any one of three separate parameters: throughput, differential pressure, and time from last backwash cycle.

This system primarily initiates the backwash cycle based on the throughput of each vessel. The differential pressure method is used to protect the system from excessive pressures in the event of heavy fouling of the media.

A summary of the backwash characteristics of the Springfield system is presented in Table 10-5.

During a backwash cycle, one vessel is backwashed at a time, immediately followed by the next vessel. The backwash waste is collected in a 60,000-gal concrete holding tank located beneath the greensand vessels. After all three vessels have been backwashed, 44,400 gal of waste is contained within the holding tank. The residuals are allowed to settle for 2 hr, which enables up to 85 percent (37,740 gal) to be decanted and recycled back to the head of the plant. The remaining 15 percent (6,660 gal) is disposed of in the sanitary sewer.

Dry-pit submersible pumps are located in a pipe gallery adjacent to the backwash holding tank. Redundant (duty/standby) pumps are provided for recycling the decant back to the head of the plant, with a separate set of redundant pumps provided for disposal to the sanitary sewer.

Table 10-5   Backwash cycle summary

Type of Vessel	Vertical, Single Cell
Number of vessels	3
Surface area per vessel (ft²)	95
Minimum filter run times (hr)	57
Backwash rate gpm/ft²	12
Backwash flow rate (gpm)	1,140
Backwash volume per vessel (gal)	14,800
Total backwash volume (gal)	44,400
Backwash holding tank volume (gal)	60,000
Recycle volume (gal)	37,740 (85%)
Disposal volume (gal)	6,660 (15%)

The decant pumps are sized to pump 115 gpm. A total of two pumps are provided, operating in a duty/standby configuration. The 115-gpm flow rate was selected based on providing 10 percent flow to the plant when operating at average capacity of 2 mgd.

Two residuals pumps also operate in a duty/standby configuration. The pumps are rated for 120 gpm, and are sized to dispose of the remaining 15 percent to the sanitary sewer over a 1-hr period. This flow rate was also closely coordinated with the sanitary sewer collection agency. NAJW learned to leave a sufficient time buffer between the end of the decant and disposal cycle and the start of the next backwash cycle to allow for the repair and correction of mechanical issues without impacting the operation of the plant.

CASE STUDY NO. 2: CITY OF TALLAHASSEE—WELL NO. 26

Introduction

The City of Tallahassee, Florida, owns and operates a water supply, distribution, and storage system that supplies water to the residents, businesses, and industries within the city. Water supply is derived from a total of 29 water supply wells that range in capacity from 200 gpm to 3,500 gpm. The northwest portion of the city water system includes 3 wells, designated as Well Nos. 19, 23, and 26, that supply that portion of the system.

The city experienced a significant number of water quality complaints in the northwest portion of the distribution system. As a result of the complaints, the city conducted extensive evaluations to determine the cause(s) of the water quality problems. The evaluations included a review of the water quality from the three supply sources of the northwest system, as well as a review of the condition of the distribution system. The pattern of water quality complaints showed that the water in the distribution system was most severely impacted when Well No. 26 was in operation. As such, the evaluations focused on water quality of Well No. 26.

Based on the water quality evaluations, the problems associated with Well No. 26 can be categorized into three components.

1. Contamination and degradation of the borehole
2. Contamination of the well discharge
3. Distribution system water quality

Although the water quality of Well No. 26 was in compliance with the primary and secondary drinking water standards, the raw water contains low levels of iron, manganese, and iron-related bacteria (IRB). Studies found that the well contained IRB in the form of *Crenothrix*, and that the levels of iron and manganese in the well were sufficient to promote growth of *Crenothrix* in the distribution system, which in turn resulted in customer complaints.

Remediation measures were undertaken to address the water quality issues associated with water quality complaints. Borehole rehabilitation of Well No. 26, involving mechanical and acid cleaning to kill IRB in the aquifer, was performed. The city also cleaned the distribution system, with pigging and disinfecting. While the borehole rehabilitation and distribution system cleaning reduced water quality complaints in the northwest service area, these measures were not considered to be a permanent solution. Additional acid treatments of the well would periodically be required, and distribution system pigging, which cannot provide complete eradication of IRB, would eventually need to be repeated because the iron in the water from Well No. 26 provides for regrowth of IRB. Another solution was needed.

Well No. 26 has a rated capacity of 2,500 gpm. The treatment plant includes unit redundancy that enables treatment of the full 2,500 gpm with one unit out of service.

Raw Water Quality Summary and Treated Water Quality Goals

Table 10-6 is a summary of the raw water quality for Well No. 26, which was in compliance with the primary and secondary drinking water standards with respect to iron and manganese. However, iron concentrations as low as 0.2 mg/L are generally considered sufficient to promote growth of IRB. Due to the known presence of IRB in the northwest district of the distribution system, the iron levels in the raw water were a water quality issue that needed to be addressed.

Although the iron and manganese levels in Well No. 26 were below the secondary drinking water standards, the primary purpose of treating the water there was to remove iron and manganese to levels that are insufficient to promote growth of *Crenothrix* in the distribution system—basically depriving the bacteria of the nutrients needed to keep them viable. Iron levels as low as 0.2 mg/L have been shown to promote growth of IRB, and manganese levels above 0.01 mg/L are believed to promote the growth of certain types of IRB, including *Crenothrix*. Thus, the treated water quality objective was to remove iron to levels below 0.05 mg/L, which is removal of iron to 75 percent below the threshold concentration that promotes IRB growth, and to remove manganese to levels below 0.01 mg/L. The northwest service area was known to respond well to lower levels of iron and manganese as evidenced by fewer water quality complaints when Well No. 26 was offline. Table 10-7 is a summary of the treated water quality goals.

Table 10-6   Raw water quality at Well No. 26

Parameter	Raw Water Concentration
Iron	0.20 mg/L
Manganese	0.05 mg/L
Total hardness	163 mg/L as $CaCO_3$
Alkalinity	148 mg/L as $CaCO_3$
Total dissolved solids	192 mg/L
Silica	11.7 mg/L
Lead	2.83 µg/L
Turbidity	0.81 ntu
Sulfate	5.22 mg/L
Temperature	70°F

Table 10-7   Treated water quality goals at Well No. 26

Parameter	Treated Water Quality
Iron	<0.05 mg/L
Manganese	<0.01 mg/L

Treatment Process

Three different treatment methods, including manganese greensand, membranes, and biological treatment, were evaluated. The manganese greensand treatment process was determined to have the lowest capital and life-cycle costs of all of the alternatives that were reviewed. The owner chose to move forward with the manganese greensand process and subsequently conducted pilot testing to confirm the performance of this option and determine the full-scale design parameters.

During the evaluation it was determined that Well No. 26 was a good candidate for the catalytic oxidation (CO) mode of manganese greensand treatment, because of the low iron and manganese levels present in the raw water. Full-scale design would proceed for the CO mode pending successful pilot testing.

Under the CO mode of operation, chlorine is continuously fed ahead of the greensand filters to aid in oxidizing the iron and manganese in the water and to continuously regenerate the greensand bed. Unlike the other modes of greensand treatment, CO does not require the feeding of any potassium permanganate, except during initial startup of the system. Instead, the manganese greensand is already preoxidized, requiring only chlorine to be fed to the raw water to keep the greensand regenerated. The filters are capped with a layer of anthracite media to remove the bulk of the precipitated iron and manganese and prevent blinding of the greensand. The pretreated water passes through the filter, at which point chlorine is again added as the final disinfectant, and subsequently discharged to the distribution system.

Pilot Studies

Two pilot studies were performed using manganese greensand in CO mode. The purpose of the first pilot study was to verify the efficiency of the greensand filtration system. The first pilot test was run at a loading rate of 3 gpm/ft^2. The purpose of the second pilot study was to observe the filter media performance at a higher loading rate and to gather information that might allow correlation between breakthrough times and lab analyses. A loading rate of 6 gpm/ft^2 was used for the second pilot test.

Both tests performed well in terms of iron and manganese removal. Table 10-8 presents a summary of the iron and manganese removal results for both studies. The levels of iron removal in both of the pilots showed that the greensand system was able to produce water that would not promote growth of IRB in the distribution system. Daily effluent sampling during Pilot Study No. 2, which was run at the higher loading rate of 6 gpm/ft^2, showed no effluent iron concentrations values above 0.04 mg/L. There was no significant iron breakthrough during any of the three filter run cycles conducted during Pilot Study No. 2. Backwashing was required due to differential pressure buildup, not deteriorating water quality.

Bacteria levels were monitored throughout both pilot tests. During Pilot Study No. 1, samples were taken of raw water, post-greensand filtrate, and point-of-entry water (water not treated by greensand pilot unit). During Pilot Study No. 2, samples were taken of the raw water and post-greensand filtrate. The data gathered during both pilot studies indicated that although there was some bacteriological activity in the raw water, there essentially was no identifiable bacteria in the chlorinated water from Well No. 26, even without greensand filtration. During the second pilot study, the spent filter backwash water was also tested for bacteriological activity. The tests did not show any significant levels of active bacteria, indicating that little buildup or growth of bacteria occurred on the media during the filter run period.

The pilot data suggested that the impact of Well No. 26 on the Northwest district was primarily from introduction of iron and manganese and not from introduction of IRB. The second pilot study showed that the greensand filter was capable of high iron removal efficiency at a loading rate of 6 gpm/ft^2, while maintaining reasonable filter run times of about 7 days.

Process Design Summary

Three horizontal, dual-cell greensand vessels, each designed to treat up to 1,250 gpm, were provided. Under normal operating conditions three vessels are in service, each treating 833.3 gpm at a liquid loading rate (LLR) of 4.5 gpm/ft^2. When one unit is out of service, the two remaining units treat 1,250 gpm at a LLR of 6 gpm/ft^2. Each vessel is designed with a straight shell of 21 ft and a 10-ft diameter. The internals of each vessel include a 12-in. anthracite cap, a 24-in. layer of GreensandPlus™ media (420/ft^3),

Table 10-8   Pilot study results summary for using manganese greensand in CO mode

Parameter	Pilot Study No. 1 (3 gpm/ft²)		Pilot Study No. 2 (6 gpm/ft²)	
	Iron (mg/L)	Manganese (mg/L)	Iron (mg/L)	Manganese (mg/L)
Raw water average	0.132	0.042	0.201	0.039
Effluent average	0.005	0.004	0.007	0.0004
Effluent maximum	0.063	0.015	0.040	0.002
Average percent removal	96%	90%	97%	99%
Duration of study	41 days		23 days	
Average filter run time	N/A		7 days	

and a 16-in. graded gravel bed. The GreensandPlus media has an effective size of 0.30 to 0.35 mm. The anthracite has an effective size of 0.6 to 0.8 mm.

Over time, the buildup of filtered particles and bed compaction in the filter cause the pressure drop across the filters to increase. When the pressure drop hits a predetermined level (10 psi maximum), the greensand vessels are backwashed in order to clean out the particle buildup, which includes the oxidized iron and manganese, and relieve the bed compaction. One vessel is backwashed at a time while the other 2 filter vessels remain on line. Each vessel has two cells that are backwashed sequentially. The purpose of the smaller operating compartments, created by dividing the vessels into two cells, is to permit a lower flow rate per vessel backwash event. The estimated filter run time between backwashes is approximately 168 hours or 7 days, based on three vessels continuously in service and a raw water quality of 0.2 mg/L iron and 0.05 mg/L manganese.

Two types of wash sequences restore the filter beds. The first is a standard backwash sequence, which primarily involves the reversal of flow through the media at a rate of 15 gpm/ft² for 10 min to expand the media and dislodge accumulated particulate matter. Effluent water from the two online filters supplies the water to backwash the third filter bed.

The second is a special backwash sequence that includes air scouring in addition to backwash. In this sequence, an additional scrubbing effect occurs when air is introduced through the bottom of the filter through the air wash distributor at a rate of 2 cfm/ft² for 10 min. This accomplishes a more thorough cleaning than the standard water backwash sequence alone. This type of backwash sequence will generally be implemented after every 5 to 15 standard backwash cycles.

In both types of backwash sequences, the reversed flow water is drained to the backwash holding tank. Before going online, all water remaining from the backwash sequences is rinsed from the bed and sent to the backwash holding tank. The standard backwash sequence lasts approximately 15 min per cell (30 min/vessel) and the backwash sequence with the air wash lasts approximately 27 min per cell (54 min/vessel). Table 10-9 describes the details of the manganese greensand system.

Residuals Handling System

A residuals handling system is part of the manganese greensand system waste. The manganese greensand residuals handling system consists of an equalization/settling tank along with pumping facilities for the decant and

Table 10-9 Manganese greensand system for treating water at Well No. 26

Vessel Details	
Number of vessels	3
Cells per vessel	2
Anthracite depth (in.)	12
GreensandPlus media depth (in.)	24
Graded gravel bed depth (in.)	16
Operating Mode	
Total flow to vessels (gpm)	2,500
Maximum flow per vessel (gpm)	1,250
Maximum flow per cell (gpm)	625
Maximum liquid loading rate (gpm/ft^2)	6.0
Filter run time (hr)	168
Maximum pressure drop (psi)	10
Backwash Mode	
Backwash flow per cell (gpm)	1,580
Backwash loading rate (gpm/ft^2)	15.0
Air flow per cell (cfm)	210
Air loading rate (cfm/ft^2)	2.0
Standard backwash duration per cell (min)	15
Special backwash duration per cell (min)	27

settled residuals. The waste wash-water volume generated from the backwash of each filter vessel is approximately 40,700 gal. This volume includes backwash water, drain-down water, and rinse-to-waste water. Each filter is backwashed approximately every 168 hr or 7 days of operation, generating a total waste washwater volume of approximately 122,000 gal/week. The waste production from this process is summarized in Table 10-10.

Backwash holding tank. Backwash water, draindown water, and rinse-to-waste water are piped directly to the backwash holding tank where the wastewater is settled for a minimum of 4 hr before decanting the supernatant to the influent of the greensand filters and subsequently disposing of the remaining settled residuals.

With its capacity of 97,000 gal, the backwash holding tank is of sufficient volume to hold the water from the backwash of 2 filter vessels (4 cells). The tank is a 34-ft-diameter, 15-ft-sidewall above-grade steel tank. The backwash holding tank parameters are summarized in Table 10-11.

Decant recycle. Generally treatment of manganese greensand residuals requires a 2-hr settling time that enables 85 percent of the residuals to be recycled. The remaining 15 percent require disposal. While the equalization tank is sized for 2 backwashes, the greensand system backwashes are intended to be sequenced so one backwash is processed at a time. Since one backwash volume is 40,650 gal, the recycle of 85 percent would mean a decanting of 34,600 gal at a time. The decant pumps discharge the supernatant at a rate of 42 gpm to the influent of the greensand filters. The decanting period for the recycled volume is approximately 14 hr. Two decant pumps are provided, one duty and one standby. A summary of the decant system is presented in Table 10-12.

Sludge disposal. After the decant period, the remaining 15 percent of the residuals (the settled portion) requires disposal. The site is adjacent to a sanitary sewer, which is used to dispose of settled residuals. Settled residuals of about 6,100 gal per backwash are pumped to the sewer at a rate

Table 10-10 Residuals handling system for manganese greensand system

Number of vessels	3
Filter run time (hr)	168
Waste volume per vessel per backwash (gal)	40,700
Total waste volume per week (gal)	122,000

of 65 gpm. The sludge disposal system has 2 pumps, one duty and one standby, and pumps sludge from the waste wash-water equalization tank to the sewer once the decanting process has been completed. The disposal of sludge takes about 1.6 hr. Settled sludge amounting to approximately 18,300 gal/week is discharged to the nearby sanitary sewer. A summary of operations for the sludge disposal system is presented in Table 10-13.

Table 10-11 Backwash holding tank specifics for manganese greensand system

Number of tanks	1
Volume (gal)	97,000
Diameter (ft)	34
Sidewall height (ft)	15
Minimum settling time (hr)	4

Table 10-12 Decant system summary for manganese greensand system

Decant volume per backwash (gal)	34,600
Number of pumps	2 (duty + standby)
Rated capacity (gpm)	42
Total dynamic head (ft)	310
Decant period per backwash (hr)	14

Table 10-13 Sludge disposal system summary for manganese greensand system

Sludge volume per backwash (gal)	6,100
Number of pumps	2 (duty + standby)
Rated capacity (gpm)	65
Total dynamic head (ft)	35
Disposal period per backwash (hr)	1.6

CASE STUDY NO. 3: RUNYON WATER TREATMENT PLANT UPGRADE

Introduction

City of Perth Amboy owns approximately 1,200 acres in Old Bridge Township, Middlesex County, New Jersey, known as the Runyon Watershed, which has served as the primary water supply source for the residents of Perth Amboy since the early 1900s. All of the city's pumped wells contain iron and manganese at levels that require treatment prior to final pumping to the distribution system.

The Perth Amboy Runyon Water Treatment Facilities have been constructed in stages over many years. The primary purpose of the water treatment facilities is to remove iron, which occurs naturally in the groundwater at concentrations averaging approximately 8 mg/L. Manganese is also present in the groundwater and is removed along with the iron in the water treatment process.

The original iron removal plant was constructed in 1926, with major improvements made in 1936. The original facility included an aeration basin, lime-feed facilities, chlorination facilities, and pressure filtration. A new plant was built adjacent to the original plant in 1971 and chemical pretreatment facilities, aerators, 2 solids-contact clarifiers, and 4 additional pressure filters were added. In 1990, construction of 4 aerators and 2 additional solids-contact clarifiers was completed.

In the early 1980s, operating difficulties at Runyon forced a portion of the new plant to be removed from service, which required the old plant to be returned to service. Raw water from Wells No. 5, 7, and 8 is discharged into the raw water sump, which is located beneath the chemical building. The water is pumped from the raw water sump to the roof of the chemical building, where aeration facilities are housed. Aeration is provided to remove carbon dioxide, to raise the raw water pH, and to add oxygen associated with oxidation of iron and manganese. The two aerators were the original 1971 units—conventional wooden-slat type with induced draft fans. These aerators were difficult to maintain and over the years became deteriorated. Figure 10-1 presents a schematic of the Perth Amboy Runyon Water Treatment Facility.

Ranney Well No. 9R discharges directly into the new aerators associated with Clarifiers No. 3 and 4, without passing through the influent sump and the raw water pumps. The four new aerators are thin film drum

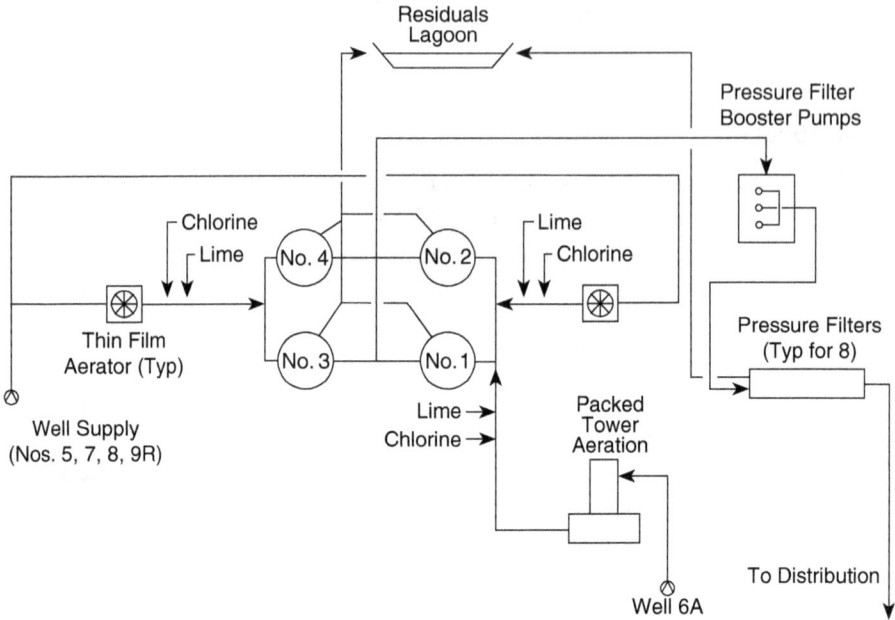

Figure 10-1 Perth Amboy Runyon Water Treatment Facility

type and were reported to operate effectively in oxidizing the dissolved iron in the water. The aerator influent piping is constructed in such a way that excess flow from Well No. 9R can be directed to Clarifiers No. 1 and No. 2 and excess flow from the raw water pumps can be directed to Clarifiers No. 3 and 4. Raw water from Well 6A is pumped to a packed tower aeration facility located adjacent to the treatment plant. Treated water from the packed tower aeration facility is pumped directly to the influent of Clarifier No. 2. Supernatant return from the backwash settling pond is pumped directly to Clarifiers No. 1 and No. 2 from the settling pond supernatant pump station.

Hydrated lime solution and 15 percent sodium hypochlorite solution are added to the aerated water as it flows by gravity from each of the aerators into the four solids-contact clarifiers. Hydrated lime solution is added to raise the pH of the water prior to its discharge into the solids-contact clarifiers. Raising the pH hastens chemical reactions that cause the iron to precipitate, forming an iron sludge. There are two hydrated-lime storage silos, each with a storage capacity of 20 tons, and two separate lime feed systems. One lime feed system is dedicated to Clarifiers Nos. 1 and 2 and the other system to Clarifiers Nos. 3 and 4. The lime is dissolved in solution tanks and pumped as a slurry into the influent of the solids-contact units with centrifugal pumps. One lime-slurry pump is dedicated to each clarifier influent. The pH of each clarifier is monitored and controlled manually.

Chlorine, in the form of 15 percent sodium hypochlorite solution, is applied to the aerated water to oxidize the iron and for disinfection. An adequate amount of sodium hypochlorite solution is added to the aerated water to satisfy the chlorine demand of the water and to produce a free chlorine residual of approximately 1 mg/L at the effluent of the pressure filters. Additional chlorine can be added to the plant finished water in the event of a low chlorine residual in the filtered water.

The four solids-contact clarifiers are individually rated for 2 mgd each and have a total combined treatment capacity of 8 mgd. The lime reacts with the iron in the solids-contact clarifiers, forming a flocculated sludge. The influent water enters the reaction zone of the clarifier, where mixing coagulation and flocculation occur. The water flows downward in the reaction zone. The flocculated water then flows upward, passing through a flocculated sludge blanket prior to entering the quiescent settling zone of the clarifier. The solids-contact clarifiers are 46 ft in diameter, with an overall height of 18 ft, a side water depth of 16.7 ft, and a surface loading rate of 0.84 gpm/ft^2 at the design flow rate of 2 mgd.

The settled water from all clarifiers is collected and flows by gravity into the clearwell located in the pressure filtration building. The water from the clearwell is pumped into pressure filters for final treatment prior to discharge into the distribution system.

Raw and Finished Water Quality and Water Quality Goals

Iron and manganese levels in the well supply and finished water of the treatment plant are summarized in Tables 10-14 and 10-15, respectively.

Table 10-14 Raw water quality data in Perth Amboy Runyon wells

Parameter	MCL or SMCL (mg/L)	Raw Water Average (mg/L)				
		Well No. 5	Well No. 6	Well No. 7	Well No. 8	Ranney Well No. 9R
Iron	0.3	4.8	13.7	10.4	13.9	4.6
Manganese	0.05	0.21	0.14	0.11	0.13	0.12

Table 10-15 Finished water quality data after treatment in Perth Amboy
Runyon Treatment Facility

Parameter	MCL or SMCL (*mg/L*)	Finished Water (*mg/L*)
Iron	0.3	<0.5
Manganese	0.05	<0.03
Color	10 CU	<8.0 CU
Surfactants	0.5	<0.3
Chloride	250	26
Copper	1.0	<0.05
Fluoride	2.0	<0.5
Hardness (as CaCO$_3$)	250	77
Sodium	50	13
Sulfate	250	57
TDS	500	151
Zinc	5	0.05
Aluminum	0.2	<0.6
Turbidity	0.50 ntu	0.36 ntu
pH	6.5–8.5	8.1
Total alkalinity	N/A	29

Runyon Water Treatment Plant Improvements

As a result of evaluations of the treatment plant process components, major improvements were recommended, including the following:

- Replacement of the induced draft aerators with four new rotary drum aerators
- Replacement of two existing clarifiers
- Replacement of lime storage and feed systems and automation of the lime system
- Addition of four new pressure filters and replacement of the existing pressure filters with new pressure filters (resulting in eight new pressure filters)

Aerator design improvements. New aerators were installed and operate by "stretching" the raw water passing over the center drum into a thin film with a large surface area exposed to the atmosphere, allowing it to be easily aerated. As the water passes through the drum, it is continuously renewed with oxygen-deficient water, which is then also aerated. The action of the water falling through the unit creates enough of an air draft to supply more oxygen than the water can dissolve. This eliminates the need for a motor-driven induced draft fan, which requires energy and maintenance. Each of the four units is capable of aerating up to 1,250 gpm (1.8 mgd) of raw water with a maximum iron content of 20 mg/L.

Solids-contact clarifiers. Two existing solids-contact clarifiers operated poorly and needed significant rehabilitation. After analysis, a recommendation was made to replace Clarifiers No. 1 and 2 to provide a long-term solution to clarification requirements. The clarifiers are a critical component of the treatment system because of the high iron concentration of the raw water. Insufficient clarification and removal of iron from the raw water would overload the pressure filters and may result in shorter filter run times and possible breakthrough of iron. Table 10-16 summarizes the design criteria for the new solids-contact clarifiers.

Pressure filter improvements. The filters are dual-media, dual-cell, horizontal pressure filters designed for the removal of iron floc carryover from the clarifiers. The estimated filter capacity is approximately 500 grains/ft^2 of filter area. Based on iron loading of 1.7 mg/L, the estimated filter run time is approximately 46 hr. Each filter operates at a flow rate of 1 mgd at a loading rate of 2 gpm/ft^2. The design loading for the proposed filters has been set at 2 gpm/ft^2, which is identical to the existing filter rates. Dual-cell filters were selected to reduce the required backwash flow rate by washing half of the filter at a time. The design parameters for the new and replacement filters are summarized in Table 10-17.

Each filter is equipped with two raw water inlet/backwash waste outlet connections and one filtered water outlet/backwash supply inlet connection. Operation of the filter is controlled with external, hydraulically operated butterfly valves. Filter operation is automatic and is controlled by the proposed plant automation system, with the ability to manually operate the filter in the event of a control system failure.

Automatically controlled butterfly valves include two raw water filter inlet valves, one filtered water outlet valve, one backwash inlet valve, two backwash waste outlet valves, and one rinse-to-waste outlet valve.

The raw water inlet and filter outlet piping includes separate manually operated butterfly valves for isolation. Flow rate through each filter during normal operation is limited to a maximum of 700 gpm (2 gpm/ft^2) by a hydraulic rate-of-flow control valve on the filter outlet line. Downstream of

Table 10-16 Solids-contact clarifiers design criteria summary

Parameter	Size/Type/Capacity
Capacity	2 mgd/each
Diameter	46 ft
Side water depth	18 ft, 8 in.
Overall height	20 ft, 8 in. to top of tank
Surface loading rate	0.836 gpm/ft^2
Loading rate in the clarification zone 5 ft below the level of the discharge weirs at design capacity	1.0 gpm/ft^2
Outlet weir type	V-notch, adjustable
Outlet weir loading rate	<10 gpm/ft of weir length
Total detention time at design capacity	2 hr, 45 min
Minimum reaction time at design capacity	10 min
Sludge withdrawal	Automatic intermittent type with manual override and motor-operated plug valve controlled via DCS system
Sludge sample taps	Manual withdrawal type
Influent pipe diameter	10 in.
Bottom	Sloped concrete

Table 10-17 Pressure filters design criteria summary

Parameter	Size/Type/Capacity
Filter size	38 ft long × 10 ft diameter
Filter loading rate	2 gpm/ft^2
Filter capacity	1 mgd
Design pressure rating	150 psi
Internal area	380 ft^2/filter
Underdrain distributor	Header-lateral w/brass and Monel sand valves common to both cells located on 15-in. centers
Inlet distributor	Header-lateral design, separate systems for each cell
Graded gravel bed depth	16 in. total
Filter media	18-in. anthracite (0.6–0.8 mm) 24-in. filter sand (0.45–0.50 mm)

the flow-control valve is an orifice plate used to control the rate-of-flow control valve. A flow transmitter that measures the filter flow rate is also connected to the orifice plate. The backwash flow rate is controlled by a Venturi flowmeter and a hydraulic flow-control valve on the backwash supply header, at a point common to all filters.

Backwashing

Backwashing the pressure filter removes the particulate material that has been captured and retained by the filter and loosens the filter bed to relieve compaction and filter head loss. The pressure drop across a filter should never exceed 10 psi. During backwash, the flow of water through the filter is reversed by opening and closing the appropriate external butterfly valves. Water will enter the underdrain distribution system in the bottom of the filter and flow upward, fluidizing the filter media and removing the oxidized iron and other foreign material accumulated in the filter bed during the service run. The filter design includes provisions for a 10-min backwash at 20 gpm/ft^2 per cell, which equates to 3,800 gpm or 5.46 mgd. After backwashing, the filter is rinsed (filtered to waste) for about 3 min before the filtered water is allowed to flow to the distribution system.

Anticipated Filter Runs

The filter run duration between backwashing cycles is a function of the iron, manganese, and other filterable constituent concentrations in the settled water from the clarifiers. The anticipated filter run length is determined by operating a filter at maximum capacity (700 gpm), using 500 grains/ft^2 removal capacity, with a design iron level of 1.7 mg/L.

Where: V = volume
t = time

500 grains/ft^2 × (1 lb/7,000 grains) × (380 ft^2/filter)
  = 27.14 lb/filter run
1.7 mg/L (Fe) × V (mil gal) × 8.34 lb/gal
  = 27.14 lb/filter run
V = 1.91 mil gal/filter run
700 gpm × t (min/filter run)
  = 1.91 mil gal/filter run
t = 2,735 min = 45.6 hr/filter run

If the actual iron concentration is less than 1.7 mg/L, the filter run time between backwashing will be proportionally greater. Likewise, if the iron concentration is more than 1.7 mg/L, the filter run time will be less.

Backwash Rates and Backwash Water Supply

For the purpose of estimating the maximum quantity of backwash supply and wastewater, it is estimated that each pressure filter will be normally backwashed for 10 minutes per cell at a rate of 20 gpm/ft^2. The rinse to waste following the backwash sequence is performed with raw water and therefore does not add to the backwash supply requirements. Therefore, the total backwash volume per filter is as follows:

$$V = (10 \text{ min/cell} \times 20 \text{ gpm/ft}^2 \times 190 \text{ ft}^2/\text{cell}) \times 2 \text{ cells}$$
$$V = 76,000 \text{ gal/filter}$$

The new clarification, aeration, and filtration systems were all constructed and have been successfully operating to produce water that meets or exceeds the finished water quality goals of the project.

CASE STUDY NO. 4: UNITED WATER—WYANDOTTE WTP

Introduction

United Water New Jersey (UWNJ) owns and operates three water supply wells located in Franklin Lakes, New Jersey. The wells are known as Wyandotte Well No. 1, Wyandotte Well No. 2, and the High Mountain Well. The two Wyandotte wells are located on an easement within a public park, adjacent to Wyandotte Drive. The High Mountain Well is located on a separate easement within the same park, adjacent to High Mountain Road. Water from the Wyandotte wells was historically treated at the Wyandotte Drive location, with a manganese sequestration chemical and sodium hypochlorite for disinfection. In accordance with the water allocation permit, the two Wyandotte wells are configured so that only one of the wells is permitted to be operated at a time.

Design Basis

The new Wyandotte treatment facility was designed to provide a firm treatment capacity of 1,150 gpm, based on the maximum allocated flow from the High Mountain Well and a single Wyandotte well. The allocated pumping rate for each well is provided in Table 10-18.

Raw Water Quality and Treated Water Goals

The treatment facility receives raw water from the High Mountain and Wyandotte wells. Raw water from the High Mountain Well is actually taken downstream of the existing ultraviolet disinfection and caustic treatment, since the High Mountain Well is under the direct influence of surface water and has a lower pH than the Wyandotte supply. Table 10-19 summarizes estimated composite water quality, using a weighted average assuming 65 percent of the combined flow originates from a Wyandotte well and 35 percent originates from the High Mountain Well. Design raw water concentrations reflect the historical maximum for each parameter except for iron and manganese. To account for possible future increases in iron and manganese concentrations and varying operating scenarios, the plant was designed for up to 0.2 mg/L of influent iron and up to 0.2 mg/L of influent manganese.

Table 10-18 Allocated pumping rate for Wyandotte and High Mountain wells

Operating Well Location	Allocated Pumping Rate (gpm)
New Wyandotte Well	750
High Mountain Well	400
Total	1,150

Table 10-19 Raw water quality at the Wyandotte treatment facility

Parameter	Wyandotte Average Raw	High Mountain Average Raw	Composite Raw Water	Design Raw Water (Max)
Iron (mg/L)	0.09	0.05	0.08	0.2
Manganese (mg/L)	0.10	0.07	0.09	0.2
Total hardness (mg/L as $CaCO_3$)	222	180	207	270
Alkalinity (mg/L as $CaCO_3$)	159	157	159	260
Total dissolved solids (mg/L)	368	400	379	500

Table 10-20 Treated water quality goals at the Wyandotte treatment facility

Parameter	Treated Water
Iron (mg/L)	0.2
Manganese (mg/L)	0.05
Total hardness (mg/L as CaCO$_3$)	250
Total dissolved solids (mg/L)	500

Table 10-20 summarizes the treated water quality requirements for the Wyandotte Well treatment system.

Treatment Process

The treatment system consists of the following unit processes:

- Manganese dioxide–coated sand filters for removal of iron and manganese
- Corrosion inhibitor/sequestrant treatment, which although no longer expected to be necessary for manganese sequestration, will be continued to maintain distribution system corrosion control
- Sodium hypochlorite treatment for preoxidation of iron and manganese and for final disinfection

Figure 10-2 presents a schematic diagram of the treatment process at the Wyandotte water treatment plant.

As described earlier, the new treatment is being provided downstream of the High Mountain ultraviolet disinfection and caustic treatment.

The design criteria for each of the unit processes are presented in the next sections.

Manganese greensand filtration. Iron and manganese removal is accomplished using pressure filtration with manganese dioxide–coated sand, such as Hungerford & Terry GreensandPlus or equivalent. The system is designed for operation in catalytic oxidation (CO) mode, which does not require feeding of potassium permanganate, except during initial startup of the system. Instead, the sand is already preoxidized and chlorine is fed to the raw water to continuously regenerate the bed. CO was selected for this application because of the low levels of iron and manganese and the higher loading rates that are possible with CO. Higher loading rates are desirable at this location due to the space constraints of the existing site.

UWNJ conducted pilot testing of the H&T GreensandPlus CO system at the Wyandotte and High Mountain wells to determine the manganese and iron removal efficiency of the system, particularly at higher filtration rates. Based on the pilot study results, water quality data, and treatment goals, the design parameters are provided in Table 10-21.

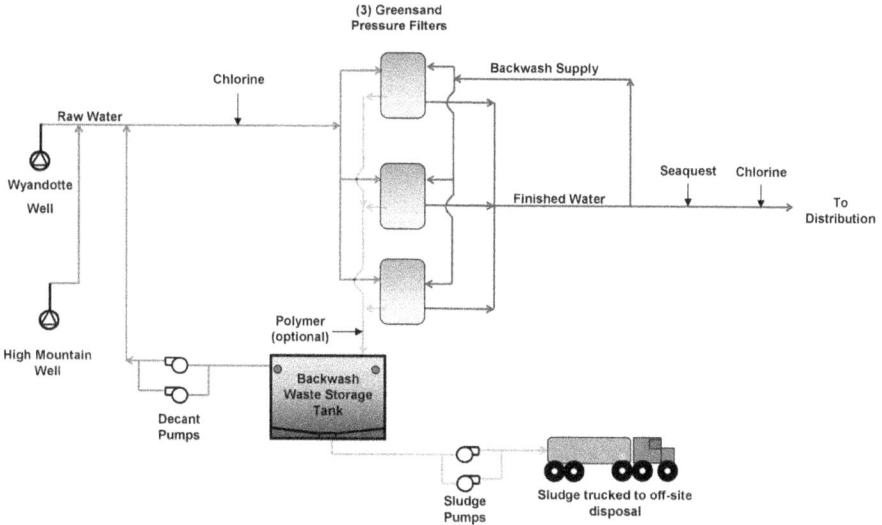

Figure 10-2 Wyandotte Water Treatment Plant

Table 10-21 Design parameters and treatment goals for Wyandotte water treatment plant

Parameter	Design Values	
	Average	Maximum
Treatment capacity	750 gpm	1,150 gpm
Number of pressure filters	3	3
Filter diameter	9 ft	9 ft
Loading rate	3.93 gpm/ft^2 (all vessels) 5.89 gpm/ft^2 (2 vessels)	6.03 gpm/ft^2 (all vessels) 9.04 gpm/ft^2 (2 vessels)
Chlorine dosage	3 mg/L maximum	3 mg/L maximum
Design raw water Mn, mg/L	0.2	0.2
Design raw water Fe, mg/L	0.2	0.2
Estimated filter run time	61 hr	40 hr

The media include a 12-in. layer of gravel, an 18-in. layer of manganese dioxide–coated sand, and an 18-in. layer of anthracite.

Residuals

Backwash wastewater from the filters is collected in an equalization/settling tank. From the holding tank, supernatant is pumped to the head of the plant. Residuals are pumped out periodically and trucked from the site since there is no local sewer. A summary of the residuals production is provided in Table 10-22.

The Wyandotte water treatment plant was commissioned and has been in service supplying high-quality water that meets all of the water quality and operating goals for the project.

Table 10-22 Residuals production

Backwash Rate	763 gpm (12 gpm/ft²) for 10 min 383 gpm rinse for 3 min
Waste backwash volume per filter	8,800 gal
Total waste backwash volume (3 filters)	26,400 gal
Decant volume per backwash (85% of total)	22,440 gal
Decant return flow rate	75 gpm (return in 5 hr)
Settled sludge volume per backwash (15%)	3,960 gal
Backwash holding tank capacity	39,000 gal

GLOSSARY

absorption Fully drawing in (as a sponge absorbs water)

acid A compound that yields positive hydrogen ions in a water solution; a strong acid is one that yields a high percentage of hydrogen ions in a water solution

adsorption Adhesion of the molecules of a gas, liquid, or dissolved substance to a surface. In water treatment, for example, manganese adsorption refers to manganese in solution (Mn^{2+}) adhering to the surface of manganese dioxide, $MnO_2(s)$, molecules.

aggressivity The tendency for corrosion to occur in a specific water

ammonia A light, colorless gas, NH_3, with an irritating and pungent odor. It normally has a chemical demand for chlorine of almost 8 times its own mass (i.e., 8:1).

anion A negatively charged ion attracted to the positive pole in electrolysis

anthracite A type of coal with characteristics that often make it desirable for potable water filtration. Material mined in Pennsylvania is not the only anthracite coal suitable for use in potable water filtration. This handbook refers to filter coal, which may or may not be anthracite. Other commonly known coal types are bituminous coal and lignite coal. Bituminous coal from some deposits can be used for filtration.

aquifer An underground layer of porous rock, gravel, or sand containing water

attrition The act or process of wearing away or grinding down by friction, or any gradual wearing or weakening, especially to the point of exhaustion

bench testing Small-scale testing of chemical reactions or treatment processes on a bench, counter, or table

biofilm The term used in microbiology to describe bacteria slimes or plugs. It is also used to describe the bacteria covering the granular filter media in a biofilter for removing iron and/or manganese. Biofilms are also found in water distribution system piping.

biofilter Granular meda (e.g., activated carbon, anthracite coal, or sand) that has developed a biofilm capable of degrading organic matter, ammonia, or both. Also known as biologically active filter.

carcinogen Any substance that causes cancer

catalyst (catalytic agent) A substance that increases the speed of a chemical reaction without itself being permanently changed

cation A positively charged ion attracted to the negative pole in electrolysis

caustic soda A common name for sodium hydroxide, NaOH, available in 25 percent and 50 percent concentrations

Celsius scale The thermometer scale on which the freezing point is 0° and the boiling point is 100° at standard pressure

chemical change A change in which one or more new substances are formed

clean-bed head loss (*See also head and head loss*) Head loss in a filter when the filter bed is clean. If the decision to backwash is based on head loss, then automatic equipment is triggered by the differential between the clean bed head loss and the dirty bed head loss.

clearwell The place where water is stored before being pumped through the distribution system. In many plants, the clearwell is a storage area under the facilities; water goes from there to enclosed storage reservoirs or water towers, then into the distribution systems. The clearwell may be the only storage reservoir in the system.

coagulant A substance that triggers formation of a soft, semisolid mass in water, to which constituents to be removed are attracted and/or trapped by adhesion; often the constituents become heavy enough to settle out. Many such coagulants aid iron and manganese removal, but the most common are alum and polymers of many types. Coagulants work by neutralizing or destabilizing the surface charge on the particulate to be removed, permitting flocculation to take place. (*See also flocculants*)

cohabit (cohabiting) Two or more sharing the same environment. For the purposes of this handbook, cohabit refers to microbiological organisms.

colloid A dispersion of particles larger than those in true solutions and smaller than those in true suspensions. In practical use, a colloid is a particle so fine it will not be filtered out by granular filter media (even by 0.30–0.35 mm manganese greensand). To filter colloids, a coagulant/flocculant first induces the particles to bunch up into units large enough to be filtered. The colloids most commonly removed from potable water carry negative surface charges.

complexing *See organic complexing.*

compound chemical A substance composed of two or more elements chemically combined in definite proportions. For example, H_2O is a compound made up of two hydrogen and one oxygen atoms.

conditioning (*See also regeneration*) Usually, a reference to the act of applying a chemical in solution to a new, unused product, such as manganese greensand, to transform the material into a condition in which it can do its job. Before the $MnO_2(s)$ coating on manganese greensand can adsorb optimum amounts of manganese in solution, it must first be conditioned or regenerated with a strong oxidant, typically potassium permanganate $(KMnO_4)$ or chlorine (Cl_2).

cost-benefit analysis A study to consider specifically both the cost and the benefit of doing something. For example, the cost to chlorinate water is $X per 1,000 m^3; the benefit is drinking water free of disease-causing organisms.

decomposition A compound breaking down into simpler substances or into its elements. Also, rot or decay.

delta P pressure differential. Usually written using the symbol for delta in the Greek alphabet, Δ, followed by a capital P: ΔP density weight per unit volume, as the grams per cubic centimeter of a solid, the milligrams per liter of a liquid, or the grams per liter of a gas

dewatering The removal of water from something. For example, in lime softening processes, large volumes of lime sludge accumulate. In order to dispose of the lime sludge, it is dewatered and then handled as a solid or semisolid by-product.

disinfection Making water safe for humans to drink, normally by adding chlorine, chlorine dioxide, or ozone, or by applying ultraviolet light

dosage The amount of chemical added to raw water during water treatment, usually expressed in milligrams per liter (mg/L). For example, 0.64 mg/L chlorine (Cl_2) helps to oxidize iron; other dosages might be stated as 1 mg/L fluoride, 5 mg/L sequestrant, etc.

eductor A device used to place filter media into a filter. The granular filter medium (e.g., coal or sand) is placed into a hopper, on the bottom tip of which a Venturi device transmits water under pressure (485–690 kPa, 70–100 psi) through a hose, creating sufficient suction to draw the filter medium out of the hopper into the hose, where it forms a slurry for delivery to the filter vessel.

effective size The sieve size such that 10 percent by weight of a granular media is smaller, often shown as e.s. or ES. For example, manganese greensand has an ES of 0.30–0.35 mm, or a mean ES of 0.325 mm. This statement means that 10 percent by weight is smaller than 0.325 mm, and 90 percent by weight is larger than 0.325 mm.

electrolysis Decomposition of a substance, melted or in solution, by application of an electric current

electrolyte A substance that conducts an electric current when melted or in solution; also, a substance that ionizes in water

electrovalence The process in which metals give up electrons and nonmetals take on electrons when they combine to form a compound

element The simplest form of matter that can be obtained by any ordinary means, such as iron, oxygen, etc.

exchange medium The common term for natural and synthetic zeolites. Although exchange media are used to remove a variety of elements and compounds from industrial process water, in potable-water treatment, the term describes the medium used to soften water using the cation exchange process. (*See also zeolite*)

filter coal As used in this handbook, a comprehensive term for all coals used in potable-water filtration

filter startup The time when water starts flowing through a filter, including the first time ever, the first time following the installation of new filter media, the start of a new filter run following backwashing, the time at which a filter automatically returns to service, or any other time the filter begins service

fines Dust and other small particles resulting from the manufacturing process for granular filter media. All filter media produced by crushing contains fines, which must be removed after installation in a filter vessel and before the filter is put into service. Particle breakdown (attrition) within the filter can also contribute to the level of fines.

finished water Water ready to enter the distribution system

flocculants A group of chemicals that create floc, composed of the chemicals and aggregates of coagulated smaller particles, that can be either settled out or removed by granular media filtration. Flocculation results from a large number of collisions between coagulated particles. (*See also coagulant*)

fluidization Bringing about the state in which a granular filter medium behaves as a fluid because a sufficient volume of backwash water flows through the filter bed. Generally, if a prodder (yardstick, copper pipe, rake, shovel handle, etc.) can be easily forced to the bottom of a filter bed during the backwash cycle, the bed is said to be fluidized. The main factor influencing the degree of fluidization is the volume of backwash flow.

formula, chemical An expression containing the symbols of the atoms and subscripts (the numbers written below the baseline) showing how many of each kind of the atoms are present in the molecule; also the abbreviation for the molecule of an element or a compound. Examples include O_2 for oxygen gas or H_2O for the water molecule.

granular filter media Sand, crushed quartz, garnet sand, manganese greensand, and filter coal. Typical sizes range from 0.25 mm to 1.20 mm, although some sizes can be larger or smaller than this range. Water treatment uses other granular filter media, but this list cites those most often used.

hard water Water containing mineral ions that react with ordinary soap to form insoluble substances. The most common hardness elements are calcium and magnesium (not to be confused with manganese). Stated simply, more softening allows water to more easily generate soap lather.

head The amount of energy possessed by a unit quantity of water at its given location, usually expressed in feet or meters

head loss A reduction in head. The source of head in a water treatment plant, i.e., the water level, is not very flexible, and therefore head loss is an important consideration. The major loss of head in a filtration system is caused by the filter media itself, since friction results from restrictions in flow. Head loss increases as these restrictions increase or the flow passages diminish in time. Head loss varies with the depth, shape and size, and cleanliness of the media. As the media becomes dirty due to accumulation of trapped particulate material (including oxidized iron and manganese), head loss grows (indicated by a rising reading on the pressure gauge on the raw water line feeding into the filter, due to an increase in pressure needed to force water through the filter). Thus, head loss is a means of determining when backwashing is required. Head loss over 2.5 m (8 ft) is generally accommodated in the design of a gravity filter.

hydraulic carrying capacity At a given pressure, the amount of water a pipe can carry before friction results in unacceptably high pump head loss. Hydraulic (water) carrying capacity can clearly decline if deposits of oxidized iron and manganese, bacterial growths, crud, silt, etc., build up in a potable-water distribution system. Such buildup reduces the inside diameter of the pipe and raises the electrical costs of driving pumps.

hypochlorite Any salt of hypochlorous acid (HOCl)

in situ In water treatment, the location of a process where it is specifically needed. For example, coating filter medium particles with $MnO_2(s)$ in the filter during operation, as compared to artificially coating particles in a factory somewhere away from the treatment plant.

inversions *See water inversion.*

ion An electrically charged atom or radical

ionization Separation (dissociation) of a molecule in a water solution into electrically charged atoms or radicals called ions

ion exchange A chemical process for reversibly transferring ions between an insoluble solid, such as a resin, and a fluid mixture, usually a water solution. The exchange in water of the ions of one element for the ions of another. For example, sodium or potassium cations attached to a zeolite (as in a water softener) exchange themselves for cations of calcium and magnesium, softening the water passing through the zeolite bed. (*See also hard water; zeolite*)

manganism A syndrome that may result from chronic exposure to high levels of manganese, characterized by feelings of weakness and lethargy, tremors, a mask-like face, and psychological disturbances

maximum contaminant level (MCL) A value defined under the Safe Drinking Water Act as the maximum permissible level (concentration) of a contaminant in water delivered to any user of a public water system. MCLs are legally enforced standards in the United States.

$MnO_x(s)$ Manganese oxides, including manganese dioxide—$MnO_2(s)$. Throughout this handbook, the formula for manganese dioxide is shown as $MnO_2(s)$ in order to be technically and chemically correct, since subtle but important differences distinguish MnO_2 from $MnO_2(s)$. However, the term manganese dioxide is considered acceptable common usage.

molecular weight The sum of the atomic weights of the atoms in a molecule. The molecular weight of a molecule compared to the carbon atom is 12, equivalent to the total numbers of protons and neutrons in the nuclei of the atoms forming the molecule. Science has determined the practical effect of molecular weight on organic complexing by certain carbon compounds, which can have an impact on designing processes for iron and manganese removal.

O_2(aq) Oxygen in an aqueous solution, i.e., oxygen in water. Most commonly, oxygen is forced into water by simple aeration.

organic chemistry The chemistry of carbon compounds. The terms organic complexing and organic binding refer to the process through which such elements as iron and manganese become part of a carbon compound. Chemists have identified literally thousands of carbon compounds.

organic complexing (*Also organically bound, organic binding, carbon complexed*) In iron and manganese removal, associations between the metals and organic carbon compounds. Complexation is the inactivation of an ion by addition of a reagent that combines with it and, in effect, prevents it from participating in other reactions.

organic compound A compound in which carbon is the chief constituent. Originally, this term meant any compound obtained from living organisms.

oxidation In a limited meaning, a chemical combination with oxygen. In a broader sense, this term means the loss of one or more electrons (negative valence) from an element or a radical or an increase in positive valence.

oxidation–reduction A chemical reaction in which one reactant is reduced (gains one or more electrons) and another is oxidized (loses one or more electrons). The process based on this reaction is referred to commonly as redox. (*See redox*)

pilot testing Evaluation, on a scale larger than laboratory scale but smaller than full scale, of the amenability of water to treatment by particular operations or processes

polymer A general term for chemicals composed of long chains of molecules of known electrical charge and electrical strength. These compounds aid water treatment by agglomerating (clumping together in bunches) very small particles so that they can settle out of water and/or become trapped in filters.

pyrolusite A mined solid particle of manganese dioxide that does the same job as the $MnO_2(s)$ coating on manganese greensand. It is much harder than the glauconite sand base of manganese greensand, so it will not break down under filter head differential pressures as low as 55 kPa (8 psi). A typical sand-pyro matrix (mixture) used in filtering contains 10 to 50 percent pyrolusite. It has been the $MnO_2(s)$ filter medium of choice for iron and manganese removal in Great Britain for decades.

radical A group of two or more elements that act as one in any chemical reaction. For example, in Na_2SO_4 (sodium sulfate), $(SO_4)^{2-}$ is a radical.

raw water Water as it comes from the source (well, lake, reservoir, river), or untreated water

redox In a groundwater system, the microbial biomass (a growth of biological organisms) often focuses at the oxidation–reduction interface at electrical potential (E_h) values from +150 to –50 mV. Since an oxidative (aerobic) state (with positive E_h values) will support aerobic microbial activities, while a reductive state (with negative E_h values) will encourage anaerobic activities, the range E_h +150 to –50 mV is the scientific description for the change from aerobic to anaerobic states.

regeneration Production of (a compound, product, etc.) again chemically, as from a derivative or by modification to a physically changed, but not chemically changed, form. In water treatment, regeneration is the process by which chemically coated filter media such as manganese greensand, which is coated with $MnO_2(s)$, or water softener zeolite resin (a cation exchange medium) is returned or restored to its top-producing condition. To regenerate manganese greensand or pyrolusite, the regenerant used is either chlorine (Cl_2) or potassium permanganate $(KMnO_4)$. Water softener zeolites are typically regenerated with a water solution of either sodium chloride or potassium chloride (both softener salts).

residual The amount remaining after a process has been completed. For example, a free chlorine residual of 0.5 mg/L is left over after all the other chlorine demands in a particular water have been satisfied. A manganese residual is the amount remaining in the water after the process has removed all it can just prior to a residual test.

Rochelle salt Potassium sodium tartrate ($NaKC_4H_4O_6$)

secondary maximum contaminant level (SMCL) A nonenforceable numerical limit set by the USEPA for a contaminant on the basis of aesthetic effects to prevent an undesirable taste, odor, or appearance that could have an adverse impact on the public welfare

sequestrant Any agent or chemical capable of separating an element out of a compound. For example, polyphosphates can separate manganese from a compound of which it is part without changing the chemical description of that compound or the manganese sequestered. The "separating" is sequestration.

solute Material dissolved in a liquid

sorption A surface phenomenon involving absorption, adsorption, or a combination of the two

specific gravity The density of a substance compared to the density of some other substance as a standard. Water is the standard for solids and liquids, while air is the standard for gases. For filter media, specific gravity (often shown as s.g. or SG) is the ratio of the mass of a volume of media to the mass of an equal volume of water under specified temperature conditions. (ASTM Standard Test C128-84 specifies this temperature as 23°C.) The SGs of many filter sands average 2.60–2.65.

static mixer Typically, a pipe with internal vanes or baffles to generate turbulence in water flowing through it to promote total mixing of a chemical with a raw-water flow. A static mixer is often the same size as the raw water line. The word static means it has no moving parts.

stratification In this handbook, one of two occurrences: (1) Filter media remaining in a layer by itself or returning to such a layer following inter-mixing with other filter media by air scouring and/or water backwashing. For example, coal topping manganese greensand should remain in a layer above manganese greensand with little intermixing where the two media meet (also known as interfacial mixing). (2) A single filter medium sepa-rating out in layers based on particle sizes and densities (specific gravities). For example, as new greensand is thoroughly backwashed for the first time, comparatively large, heavy particles move to the bottom of the greensand layer, while the undesirable fine particles move to the top of the layer, from which they can be manually removed.

temporary hardness Equivalent to alkalinity, if alkalinity is less than total hardness

total dissolved solids (TDS) The mass of ions plus silica

TPTZ 2,4,6-tripyridyl-s-triazine, a chemical used as a reagent for iron

transpiration The method by which plants take moisture from the soil and transmit it to the atmosphere

turbidity Solid particles in a given volume of water. Most turbidity meters interpret particle densities as nephelometric turbidity units (ntu). Water with a high turbidity value has a cloudy or unclear appearance; turbidity of drinking water should not exceed 1.0 ntu.

uniformity coefficient A dimensionless factor to describe the uniformity of particle size in a granular filter medium. It is defined as the sieve size that passes 60 percent of the media grains by weight divided by the sieve size that passes 10 percent of the media grains by weight. Also shown as u.c., UC, and sometimes U_C. A UC of 1 would mean that all particles are exactly the same size (a situation not possible through commercial pro-cesses). Uniform filter sand has a UC of 1.3 or less, while a UC of 1.7 is considered lacking in uniformity.

valence The number of electrons that an atom or radical can lose, gain, or share with other atoms or radicals. Valence means essentially combining power; it is the relative worth of an atom of an element in combining with the atoms of other elements to form compounds. The valence number of an element is the number of its electrons associated with formation of a particular compound. For example, the valence of Mn^{+6} is 6.

viscosity A term from physics that describes a liquid's thickness. Formally, it is the internal friction of a fluid—caused by molecular attraction—that makes it resist flowing. The viscosity of water is influenced by temperature. Water is at its maximum viscosity (its thickest or maximum density) at 4.1°C (+39.4°F). Temperature affects the viscosity of water to such an extent that filter backwash rates routinely take into account the water temperature.

water inversion The turnover of water in impoundments caused by temperature changes. For example, when surface water cools as the weather gets colder, the water at the bottom of the reservoir remains warmed by the earth. When the surface water goes down to 4.1°C (39.4°F), that layer of water reaches its maximum density and slowly begins to sink, displacing the water at the bottom, which is forced to the top of the reservoir. If the water at the bottom of the reservoir is anaerobic (i.e., lacking oxygen), it will have iron and manganese in solution, which it will carry to the surface. There, mixing with oxygen from the atmosphere by wind and wave action slowly oxidizes the iron and manganese, forming solid precipitates that fall to the bottom of the reservoir again.

zeolite Any of a large group of natural hydrous aluminum silicates of sodium, calcium, potassium, or barium, chiefly found in cavities of igneous rocks and characterized by a ready loss or gain of water by hydration; many are capable of ion exchange with solutions.

References

Introduction

CBC News. September 13, 2013. "Winnipeg's Brown Water Poses Risk, Biologist Says."

Chapter 1

Bean, E.L. 1962. Progress report on water quality criteria, *Journal AWWA*, 54(11): 1313–1331.

Cooke, Rosa-lee. 2014. Lesson 2: Health effects of iron and manganese in drinking water. Big Stone Gap, VA: Mountain Empire Community College.

Cullimore, D. Roy. 1993. *Practical Manual of Groundwater Microbiology*. Chelsea, MI: Lewis Publishers.

Environment Canada. 1994. *A Primer on Fresh Water, Canada*. Ottawa, ON: Minister of the Environment.

Freeland-Graves, J.H. et al. 1987. Manganese requirements of humans, *Nutritional Bioavailability of Manganese*. Washington, DC: American Chemical Society.

Greger, J.L. 1999. Nutrition versus toxicology of manganese in humans: Evaluation of potential biomarkers, *Neurotoxicology*, 20: 205–212.

Health Canada. 1987. Manganese. Ottawa, ON: Health Canada.

Joint FAO/WHO Expert Committee on Food Additives (JECFA). 1983. *Evaluation of Certain Food Additives and Contaminants*. Geneva: World Health Organization.

Schroeder, H.A. et al., 1966. Essential trace metals in man: manganese, *J. Chron. Dis.* 19:545–571.

USEPA. 2003. Health Effects Support Document for Manganese. PA 822-R-03-003. Office of Water (4304T) Health and Ecological Criteria Division. Washington, DC. www2.epa.gov/sites/production/

files/2014-09/documents/support_cc1_magnese_healtheffects_0.pdf. Accessed 8/24/2015.

USEPA. 2010. *Manganese Compounds Hazard Summary.*

WHO (World Health Organization). 2003. Iron in Drinking-water: Background document for development of WHO Guidelines for Drinking-water Quality. WHO/SDE/WSH/03.04/08.

WHO. 2011. Manganese in Drinking-water: Background document for development of WHO Guidelines for Drinking-water Quality. WHO/SDE/WSH/03.04/104/Rev/1.

Chapter 2

Carnegie Mellon. 2015. College of Engineering Center for Sustainable Engineering Overview. http://engineering.cmu.edu/research/centers/cse.html. Accessed 4/21/2015.

CSE (Center for Sustainable Engineering). 2015. The Center overview. http://www.csengin.org/csengine/index.html?skuvar=135. Accessed 4/21/2015.

Ford Jr., William Clay. 2010. Verbatim: How businesses view sustainability & CSR reporting, *Business Ethics*, July 27, 2010.

Intergovernmental Panel on Climate Change. 2014. Climate Change 2014 Synthesis Report: Summary for Policymakers. New York City: Cambridge University Press.

LEED (Leadership in Energy & Environmental Design). 2014. LEED Stands for Green Building Leadership. US Green Building Council. http://www.usgbc.org/leed.

RIT (Rochester Institute of Technology). 2015. The Kate Gleason College of Engineering Masters of Engineering Programs Sustainable Engineering overview. http://www.rit.edu/kgcoe/program/sustainable-engineering-0. Accessed 4/21/2015.

United Nations World Commission on Environment and Development. 1987. Report of the World Commission on Environment and Development: Our Common Future (Brundtland Commission Report). http://www.un-documents.net/our-common-future.pdf. Accessed 4/21/2015.

USEPA. 2014. What Is Sustainability? http://www.epa.gov/sustainability/basicinfo.htm#sustainability.

Chapter 5

Knocke, W.R., J.E. VanBenschoten, M. Kearney, A. Soborski, and D.A. Reckhow. 1990. *Alternative Oxidants for the Removal of Soluble Iron and Manganese.* Denver: AWWA Research Foundation and AWWA.

Chapter 7

Birm Specification Sheet. 1988. Clack Corporation, Wisconsin.

Knocke, W.R., S.C. Occiano, and R. Hungate, 1991. Removal of soluble manganese by oxide-coated filter media, *Journal AWWA* 83(8): 64–69.

Mouchet, Pierre. 1992. From conventional to biological removal of iron and manganese in France. *Jour. AWWA,* 84(4): 158–167.

Voorinen, A. et al. 1988. Chemical, mineralogical, and microbiological factors affecting the precipitation of Fe and Mn from groundwater. *Water Science & Technology,* 20(3): 249.

Chapter 8

Metcalf & Eddy: AECOM. 2013. *Wastewater Engineering: Treatment and Resource Recovery,* 5th ed. George Tchobanoglous, Franklin Burton, H. David Stensel, editors. New York: McGraw Hill.

US Environmental Protection Agency. 2004. Method 9095B, Paint Filter Liquids Test.

Chapter 9

Appenzeller, B.M.R., M. Batté, L. Mathieu, J.C. Block, V. Lahoussine, J. Cavard, and D. Gatel. 2001. Effect of adding phosphate to drinking water on bacterial growth in slightly and highly corroded pipes. *Water Res.,* 35(4): 1100–1105.

American Water Works Research Foundation and DVGW-TZW. 1996. *Internal Corrosion of Water Distribution Systems,* 2nd Edition. AwwaRF and VGW Technologiezentrum Wasser.

Clement, J.A., et al. 2002. *Development of Red Water Control Strategies,* Denver: Water Research Foundation.

Larson, T. 1975. *Corrosion by Domestic Waters,* Illinois State Water Survey Bulletin 59. Urbana: Illinois State Water Survey.

Lytle, D.A., P. Sarin, and V.L. Snoeyink. 1003. The effect of chloride and orthophosphate on the release of iron from drinking water distribution system cast iron pipe. In *Proc. of the AWWA Water Quality and Technology Conference.* Denver, CO: AWWA.

McNeill, L.S. and M. Edwards. 2001. Iron pipe corrosion in distribution systems, *Journal AWWA*, 93(7): 88–100.

McNeill, L.S. and M. Edwards. 2002. Phosphate inhibitor use at U.S. utilities, *Journal AWWA*, 94(7): 57–63.

NO-DES. 2015. Website. http://www.no-des.com/benefits-of-the-no-des-flushing-method/

Sarin, P., V.L. Snoeyink, J. Bebee, W.M. Kriven, and J.A. Clement. 2001. Physico-chemical characteristics of corrosion scales in old iron pipes, *Water Res.* 35:12:2961–2969.

Sarin, P., J.A. Clement, V.L. Snoeyink, and W.M. Kriven. 2003. Iron release from corroded, unlined cast-iron pipe, *Journal AWWA*, 95(11): 85–96.

Sarin, P., V.L. Snoeyink, D.A. Lytle, and W.M. Kriven. W. 2004a. Iron corrosion scales: model for scale growth, iron release, and colored water formation, *Water Res.* 38:1259–1269.

Sarin, P., V.L. Snoeyink, D.A. Lytle, and W.M. Kriven. 2004b. Iron release from corroded iron pipes in drinking water distribution systems: effect of dissolved oxygen, *Jour. Environ. Eng.* 130:364–373.

Utility Service Group, 2014. Case Study: Pressure Cleaning, Coating, & Repair in Avon, Connecticut.

ADDITIONAL RESOURCES

ASCE and AWWA. 1996. *Technology Transfer Handbook: Management of Water Treatment Plant Residuals.* Washington, DC: United States Environmental Protection Agency.

American Water Works Association (AWWA). 2011. *Water Quality & Treatment, A Handbook on Drinking Water*, 6th ed. James K. Edzwald, editor. New York: McGraw Hill.

AWWA. 2010. Principles and Practices of Water Supply Operations: *Water Treatment*, 4th ed. Denver, CO: AWWA.

AWWA. 2012. Process Residuals, chapter 18 of *Water Treatment Plant Design*, 5th ed. Stephen J. Randtke and Michael B. Horsley, editors. New York: McGraw Hill.

Crits, George. 2012. *Crits Notes on Water and Ion Exchange.* Revere, MA: Chemical Publishing.

Crittenden, John C., R. Rhodes Trussell, David W. Hand, Kerry J. Howe, and George Tchobanoglous. 2012. *MWH's Water Treatment Principles and Design,* 3rd ed. New York: John Wiley & Sons, Inc.

Pizzi, Nicholas G. 2010. *Water Treatment Plant Residuals Pocket Field Guide.* Denver, CO: AWWA.

Pizzi, Nicholas, and William C. Lauer. 2013. *Water Treatment Operator Training Handbook*, 3rd ed. Denver, CO: AWWA.

Rompré, A., et al. 2000. Impact of biodegradable organic material on fixed and suspended biomass, *Proceedings of the AWWA Water Quality Technology Conference*, New Orleans. Denver: AWWA.

Sarai, Darshan Singh. 2005. *Basic Chemistry for Water and Wastewater Operators*, revised edition. Denver, CO: AWWA.

Appendix A
Pilot Studies

Purpose and Need

Water treatment design and operating professionals recognize the importance of establishing the optimum design criteria for the facility. Pilot testing is generally needed to evaluate the following:

- Innovative technologies for which the regulatory agencies have no approved design criteria
- Loading rates above industry design standards
- Innovative oxidants/pretreatment chemicals
- Surface water with variable water quality
- Groundwater sources subject to chemical contamination beyond iron and manganese (such as radionuclides, arsenic, VOCs)

For larger water systems, pilot testing can be performed as part of the initial planning phase to assist in the comparison of treatment alternatives. For larger facilities, the savings associated with optimizing the selection of the best alternative can compensate for the cost of the pilot testing.

For smaller groundwater systems, pilot testing is often performed after the optimum treatment system is selected. The basic purpose of a pilot study for these systems is to establish a process for removing iron and manganese below target levels. However, pilot studies go well beyond this basic purpose to perform other functions, including:

- Determining a site-specific chemical treatment program. The study usually provides data for calculating yearly volumes of chemical required and estimating annual chemical costs.
- Specifying hardware items (e.g., chemical pumps, valves, instrumentation, air blowers, and underdrains)

- Specifying filter media
- Developing data for calculating filter run time
- Developing site-specific backwash and air-scour rates and procedures
- Determining residuals generation volumes and residuals water quality to evaluate handling and disposal options
- Providing information for residuals-handling options, such as spent-filter backwash settling rates and detention time, concentration of thickened solids, and decant concentration

Studies undertaken before initiating plant design provide the design engineer with data about appropriate filtration rates, required detention times, aeration requirements if any, sequence of chemical addition, and other prime design data. The pilot testing also allows the owner and operators to gain familiarity with the proposed equipment. This site-specific data will improve the quality of the design documents.

The pilot-testing program should begin with the development of a pilot-testing protocol that describes the entire testing program. The protocol should identify the objectives of the pilot program and include the following components:

- *Equipment description.* A process and instrumentation diagram should be developed that includes all online analyzers as well as all manual sampling points. An equipment layout should be provided that provides dimensions of the equipment.
- *Analytical description.* The water quality testing should be comprehensively described in terms of the following:
  - Location of sample points
  - Parameters to be analyzed
  - Frequency of analysis
  - Whether the analysis is from online instruments, whether it is to be taken as a grab sample, and whether the analysis will be performed in the field or by a laboratory
- *Duration of the pilot testing.* For groundwaters that have stable water quality, testing for 2 to 4 weeks may be sufficient. For surface waters with seasonal changes in water quality, testing for 6 to 9 months may be needed.
- *Staffing.* Many pilot plants are equipped with automation to allow operation with minimal operator involvement. The level of

automation should be established during the planning and pilot-testing phase. At a minimum, daily inspection of the pilot plant and manual grab sampling to supplement the online instrumentation should be performed.

- *Regulatory involvement.* Prior to embarking on a pilot-testing program, involvement with the regulators is recommended. Consideration should be given to submitting the pilot-testing protocol to the regulators for their review.

- *Cost.* The pilot-testing program costs will generally consist of the price of renting and installing the pilot equipment, staffing, analyses of the testing, and residuals disposal. Some pilot equipment is available in a trailer, and power and piping will need to be connected to the pilot equipment. A general contractor may need to be engaged to perform the electrical and plumbing work. For non-trailer-mounted equipment, it may be possible to locate the equipment in an existing facility. If this is not feasible, a temporary enclosure will need to be provided.

- *Residuals.* The pilot system will generate wastes that require disposal. Often residuals are disposed of in a sewer, if available. Prior to disposal of residuals in a sewer, the local sewer authority should be contacted and any concerns of the authority addressed. If a sewer is not available, it may be feasible to store the wastes in a temporary tank to be disposed of off site.

Pilot studies help conserve both water and money. Data developed in dozens of pilot studies indicate that most particulate removal occurs during the first 2 to 4 min of a hydraulic backwash. Many plants waste untold gallons of treated water in unnecessarily long backwashes. The same pilot studies have proven the value of air scour in loosening filtered particulate from the filter media, to be carried to the waste line with low flows of backwash water. Optimized backwash procedures like these save both water and money.

A pilot study conserves money in obvious ways, as shown by many examples of plants built or rehabilitated without this precaution. Design deficiencies resulting from the absence of pilot testing data include inadequate detention time, unacceptably low backwash rates, filtration rates too high to maintain, selection of filter underdrain designs based on lowest cost rather than hydraulic performance, stand-alone air-scour grids prone to plugging, chemical pretreatment processes based on incomplete data, and many others.

Before an actual pilot study begins, a study of the raw water must be undertaken. Waters with very similar characteristics can require distinctly different treatments.

Taking Water Samples

Samples for analysis in a pilot study should be collected carefully to ensure the most representative sample possible is obtained. In general, take samples near the center of a vessel or duct and below the surface. Use only clean containers (bottles or beakers) for collecting samples. Since nitric acid is often used to clean glassware, containers should be rinsed several times with the water to be tested before a sample is taken.

Collect any sample as close as possible to the source of the supply to minimize the effects of distribution system conditions. Allow the water to run for sufficient time to flush the system, and slowly fill the sample container with a gentle stream to avoid turbulence and air bubbles. Collect water samples from a well after the pump has run long enough to deliver water representative of the groundwater feeding the well.

It is difficult to obtain a truly representative surface water sample. Best results can be obtained by running a series of tests with samples taken from several locations and depths at different times of the year to represent the seasonal variation in water quality.

Samples intended to measure soluble iron in well water should ideally be taken without allowing contact with oxygen in the air (a difficult task) and certainly not from a turbulent stream or bubbling water. Analysis should be completed as soon as possible after the sample is taken to minimize contact time with oxygen. In some cases, iron is oxidized almost instantly by contact with oxygen. If the sample will be transported and analysis delayed, take care to let the well stream gently flow into the sample container, fill the container to overflowing, and seal with a tight cap. This collection method keeps oxidation to a minimum.

Depending on the nature of testing to be conducted, other special precautions in handling samples also may be necessary to prevent natural interferences such as organic growth or loss or gain of dissolved gases. For example, samples to be sent to an independent laboratory for organic carbon and ammonia nitrogen analysis need to be preserved with sulfuric acid, and samples for hydrogen sulfide analysis must be preserved with sodium hydroxide (NaOH) and ascorbic acid.

A good source of information on collecting and analyzing samples is the book *Standard Methods for the Examination of Water and Wastewater* (APHA, AWWA, and WEF 2012).

Historical Data Requirements

Regulatory agencies require the finished water to be sampled on a periodic basis for primary and secondary contaminants. Source water monitoring, however, is generally the responsibility of the individual utility. Because the efficiency of the treatment process is directly related to the source water quality, utilities should consider establishing a routine source water quality monitoring program.

The source water quality analyses should be carefully reviewed for any abnormal levels of chemicals and elements that may affect the treatment process about to undergo pilot testing. Tests should also check for interference with certain analytical procedures. For example, the Hach PAN method for determining low-range manganese concentrations (i.e., up to 0.7 mg/L) can be disrupted by certain levels of aluminum, cadmium, calcium, cobalt, copper, iron, lead, magnesium, nickel, and/or zinc.

General chemical analysis data often include much of the information necessary to calculate a Langelier saturation index (LSI) and aggressivity index. The LSI, a measure of a solution's ability to dissolve or deposit calcium carbonate scale, is often used as an indicator of water's corrosivity. The index is not related directly to corrosion, but rather to the deposition of a calcium carbonate film or scale. When no protective scale is formed, water is considered to be aggressive, and corrosion can occur.

A pilot study reviews the LSI to evaluate the need for measures to prevent calcium carbonate from covering media particles, which could change their effective size over time and/or blind off MnO_2 sites, preventing adsorption of iron and manganese.

Values needed to do an LSI calculation may be available from a general chemical analysis. They include total dissolved solids, calcium hardness, and total alkalinity. In addition, water temperature and pH values are required, but these data are obtained during the pilot study.

A general chemical analysis may also include levels of nitrates, sulfates, and sodium. Sulfates may be important, for example, because they can cause taste problems.

Chemical Data Developed On Site

Values for iron and manganese in the raw water are developed on site during the pilot study. Color, temperature, and pH should likewise be determined. Because no method provides reliable forecasts of occurrences within an aquifer, these five values (iron, manganese, pH, color, and temperature) should be measured several times during the course of a pilot study. Many iron and manganese measurements can be taken during a filter run. These levels may fluctuate, with dramatic variations in some well waters. In such cases the treatment process selected based on pilot study data must compensate for fluctuations in raw water chemistry.

Data Collected During a Pilot Study

1. *Influent and effluent quality.* The first results of a pilot study indicate initial removal efficiency. One of the main functions of a pilot study is to indicate both which processes work and which do not.

2. *Filtration performance as a function of depth.* A pilot study looks for very important information about where the filtration takes place in a filter. If much of the filtration takes place in the top regions of the filter bed, perhaps the design has specified too fine a filter medium. An adsorption process should ideally keep iron out of the $MnO_2(s)$ layer to prevent coating of the particles with iron floc and development of head losses that could reduce filter run lengths. To evaluate this possibility, data about iron and manganese penetration must be developed based on time and depth.

3. *Head loss during a filter run.* For either an open-top or pressure filter, a pilot study must measure head loss during filter runs. For an open-top gravity filter, this information indicates the length of the filtration cycle based on available driving head (i.e., the capacity to push water through a filter without overflowing the filter compartment). For a pressure filter using manganese greensand, a study must determine the length of time needed to accumulate a pressure differential of 55 kPa (8 psi). Pressure differentials in excess of this level can result in fracturing of greensand grains and loss of filter performance over time. For pressure filters not using manganese greensand, a study must determine the length of time required for the filter bed to reach its solids-holding capacity; at this point channeling commences, indicated by iron and manganese breakthrough (i.e., rapidly rising residual values).

4. *Filter ripening time.* Filter ripening is defined as the improvement in water quality from the start of a filter run until optimum quality is achieved (whatever parameters determine this quality level in a specific water treatment plant). Ripening time varies from filter to filter depending on many factors, such as nature of the influent raw water, filter flow rate, filter media design, chemical pretreatment program, and other factors. To achieve the objective of distributing only water of optimum quality, filtered water should run to waste from the beginning of a filtration cycle until optimum quality filter effluent results. If a filter is designed to start-stop-start a number of times between backwash cycles based on clearwell levels, the length of filter ripening each time can be established in a pilot study.

5. *Filter media design.* Several filter media designs can be tested if required, to prove or disprove performance expectations. A media design that performs well in one plant may not be suitable for another, despite similarities in raw waters and chemical pretreatment programs. Pilot testing of media designs also provides an opportunity to evaluate preferences.

6. *Filter loading rate.* This term describes the maximum filtration flow rate at which optimum-quality water can be produced, holding constant all other variables. Higher loading rates result in a smaller and less expensive facility. Evaluation of higher loading rates may also allow an existing plant to increase its capacity without adding more filters and without compromising water quality.

7. *Establishing backwash procedures.* The right design for the removal process (i.e., chemical pretreatment, media bed, filtration rate, etc.) is of little value if the process cannot be sustained. Deteriorating water quality, declining filter run lengths, and increasing need for backwash water may indicate failure of the underdrain to distribute backwash water evenly, leaving unacceptably high levels of filtered particulate in parts of the media bed. These problems may also result if backwash flow rates are either too low or too high, and/or the wrong backwash procedure is used. At the end of each pilot filter run, empirical data are developed. Accumulating data guide development of a procedure to optimize bed cleaning using the minimum amount of time, backwash water, and air scour.

8. *Impact of the chemical program on filter media.* Pilot testing with filter columns made of clear acrylic allows visual observation of filter media agglomeration, which is valuable in establishing an effective media bed cleaning procedure. The behavior attributed to pretreatment chemicals could prompt rethinking of a proposed media bed design. For example, the use of a certain polymer may require a deeper or coarser coal layer. Conversely, testing of a particular media combination may indicate particle agglomeration at the interface of the coal and the layer below, indicating the top coal layer is too coarse and/or too shallow for effective filtering.

9. *Optimizing pretreatment chemicals.* Jar testing is the most common method used to arrive at an optimum chemical dosage. While such a test can indicate whether the chemical chosen will form a suitable floc, it cannot predict the filterability of the developed floc. Visual observation of a chemical reaction of this nature can too easily be influenced by personal biases, which lead to inappropriate conclusions concerning optimum feed rates. Instead, measurement of both dosages and effluent quality can lead toward an optimal pretreatment program.

Appendix B
Calculation Tools

Table B-1   General conversion table

Multiply Number of	By	To Obtain Number of
Acres	0.4045	hectares (ha)
centimeters (cm)	0.0328	feet (ft)
centimeters	0.3937	inches (in.)
centimeters	0.0100	meters (m)
centimeters	10	millimeters (mm)
cubic feet (ft^3)	1,728	cubic inches (in.^{3})
cubic feet (ft^3)	0.0283	cubic meters (m^3)
cubic feet (ft^3)	0.0370	cubic yards (yd^3)
cubic feet (ft^3)	7.4810	US gallons (gal)
cubic feet (ft^3)	6.2280	Imperial gallons
cubic feet (ft^3)	28.32	liters (L)
cubic feet of water (at 60°F)	62.37	pounds (lb)
cubic feet/minute (ft^3/min or cfm)	472.0	cubic centimeters/second (cm^3/sec)
cubic feet/minute (ft^3/min or cfm)	4.72×10^{-4}	cubic meters/second (m^3/sec)
cubic feet/minute (ft^3/min or cfm)	0.0283	cubic meters/minute (m^3/min)
cubic feet/minute (ft^3/min or cfm)	0.1247	US gallons per second (gal/sec)
cubic feet /minute (ft^3/min or cfm)	0.4719	liters/second (L/sec)

(continued)

Table B-1   General conversion table (continued)

Multiply Number of	By	To Obtain Number of
cubic feet/minute/foot^2 (ft^3/min/ft^2 or fm)cfm/ft^2)	5.08	liters/second/square meter (L/sec/m^2)
cubic feet/minute/foot^2 (ft^3/min/ft^2 or cfm/ft^2)	0.3048	cubic meters/minute/ square meter (cm/min/m^2)
cubic feet/second (ft^3/sec)	0.0283	cubic meters/second (m^3/sec)
cubic inches (in.^{3})	16.39	cubic centimeters (cm^3)
cubic inches (in.^{3})	0.0164	liters (L)
cubic meters (m^3)	35.313	cubic feet (ft^3)
cubic meters (m^3)	264.2	US gallons (gal)
cubic meters (m^3)	220.0	Imperial gallons (Imp gal)
cubic meters (m^3)	10^3	liters (L)
feet (ft)	30.48	centimeters (cm)
feet (ft)	0.3048	meters (m)
feet of water	0.4335	pounds per square inch (lb/in.^{2})
feet of water	3.0	kilopascals (kPa)
gallons, Imperial	1.2009	US gallons (gal)
gallons, US (gal)	0.8327	Imperial gallons (Imp gal)
gallons, US (gal)	0.1337	cubic feet (ft^3)
gallons, Imperial (Imp gal)	3.785×10^{-3}	cubic feet (ft^3)
gallons, US gallons, Imperial	$3,334.546 \times 10^{-3}$	cubic meters (m^3)
gallons, US	3.785	liters (L)
gallons, Imperial	4.546	liters (L)
gallons, US/minute/square foot (gal/min/ft^2)	2.4476	meters per hour (m/hr)
gallons, Imperial per minute (Imp gal/min)	0.0758	liters per second (L/sec)
grains	0.0648	grams (g)
grams (g)	0.0353	ounces (oz)
grams (g)	2.205×10^{-3}	pounds
hectares (ha)	2.47	acres
inches (in.)	2.540	centimeters (cm)

(continued)

Table B-1 General conversion table (continued)

Multiply Number of	By	To Obtain Number of
inches of water	5.204	pounds per square foot (lb/ft^2)
inches of water	0.0361	pounds per square inch (lb/in.^{2})
kilograms (kg)	2.2046	pounds (lb)
kilograms/square meter	1.422×10^{-3}	pounds per square inch
kilopascals (kPa)	0.1450	pounds per square inch (lb/in.^{2})
liters (L)	0.0353	cubic feet (ft^3)
liters (L)	10^{-3}	cubic meters (m^3)
liters (L)	0.2642	US gallons (gal)
liters (L)	0.22	Imperial gallons (Imp gal)
liters/minute (L/min)	0.2642	US gallons per minute (gal/min)
liters/minute (L/min)	0.22	Imperial gallons per minute (Imp gal/min)
meters of water	9.8	kilopascals (kPa)
meters (m)	3.2808	feet (ft)
meters (m)	39.37	inches (in.)
meters (m)	1.0936	yards
square meters (m^2)	10.764	square feet (ft^2)
meters per minute (m/min)	0.0547	feet per second (ft/sec)
meters per second (m/sec)	3.2808	feet per second (ft/sec)
microns	3.937×10^{-5}	inches
milligrams/liter (mg/L)	1	parts per million (ppm)
ounces (oz)	28.35	grams (g)
parts per million (ppm)	0.0584	grains per US gallon grains/gal
parts per million (ppm)	0.7016	grains per Imperial gallon
parts per million (ppm)	8.345	pounds/million US gallons (lb/mil gal)
pounds (lb)	7,000	grains
pounds (lb)	453.6	grams (g)
pounds per square foot (lb/ft^2)	0.0479	kilopascals (kPa)

(continued)

Table B-1   General conversion table (continued)

Multiply Number of	By	To Obtain Number of
pounds per square inch (psi)	2.307	feet of water
pounds per square inch (lb/in.^{2})	0.0703	kilograms/square centimeter (kg/cm^2)
pounds per square inch (lb/in.^{2})	703.1	kilograms/square meter (kg/m^2)
pounds per square inch (lb/in.^{2})	6.895×10^3	pascals (Pa)
pounds per square inch (lb/in.^{2})	6.895	kilopascals (kPa)
square feet (ft^2)	144	square inches (in.^{2})
square feet (ft^2)	0.0929	square meters (m^2)
square inches (in.^{2})	6.45	square centimeters (cm^2)
square yards (yd^2)	0.836	square meters (m^2)
square miles (mi^2)	2.59	square kilometers (km^2)
temperature (°C) + 17.8	1.8	temperature (°F)
temperature (°F) − 32	5/9	temperature (°C)
yards (yd)	91.44	centimeters (cm)

Table B-2   Additional names and symbols

Name	Symbol
degrees Celsius	°C
kilowatt	kW
micrograms per liter	µg/L
milligrams	mg
milliliters	mL
milligrams per liter	mg/L
millimeters	mm
pascals (Pa = N/m^2)	Pa
Newtons	N

Where:
A = Area, A1 = Surface area or solid
V = Volume, C = Circumference

Rectangle
A = W × L

Parallelogram
A = W × L

Trapezoid
$A = H \times \dfrac{L_1 + L_2}{2}$

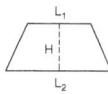

Circle
A = 3.142 × R × R
C = 3.142 × D
R = D/2
D = 2 × R

Sector Circle
$A = \dfrac{3.142 \times R \times R}{360}$

L = 0.01745 × R × α

$a = \dfrac{L}{0.01745 \times R}$

$R = \dfrac{L}{0.01745 \times \alpha}$

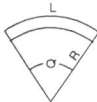

Ellipse
A = 3.142 × A × B
$C = 6.283 \times \sqrt{\dfrac{(A^2 + B^2)}{\sqrt{2}}}$

Rectangular Solid
A1 = 2[W × L + L × H + H × W]
V = W × L × H

Cone
Lateral A = ½ (perimeter of base) × S
V = 1.047 × R × R × H or
V = ⅓ (area of base) × H

Cylinder
A1 = 6.283 × R × R × H + 6.283 × R × R
V = 3.142 × R × R × H

Ellipse Tanks
$A = 6.283 \times \dfrac{\sqrt{(A^2 + B^2)}}{\sqrt{2}} \times H + 6.283 \times A \times B$

V = 3.142 × A × B × H

Sphere
A = 12.56 × R × R
V = 4.188 × R × R × R

$\dfrac{V}{231}$

For above containers:
Capacity in US gallons = when V is in cubic inches
Capacity in US gallons = 7.48 × V when V is in cubic feet

Figure B-1   Geometric formulas
Source: Anthratech Western Inc.

Index

NOTE: *f* indicates a figure; *t* indicates a table

downstream, 71
fundamentals of, 74–77, 79–83
process of, 74, 121
residuals from, 119 f, 120, 126
sustainability and, 84, 132
Clarification systems, 3, 73, 83, 115,
    126, 150 f
    design loading for, 74, 168 t
    energy requirements of, 84
    types of, 75–76
Clarifiers, 73, 77, 119, 120, 161–166
    blowdown of, 125
    construction materials for, 84
    conventional, 74, 128
    pH of, 163
    residuals and, 121, 123, 124, 126
    settling rates and, 74
    solids-contact, 3, 38, 76, 80, 81 f,
        119, 121, 150 f, 167
    types of, 119
Clean-bed head loss, defined, 176. *See
    also* Head loss
Clearwell, 58, 87, 114, 165
    defined, 176
ClO_2. *See* Chlorine dioxide
CO. *See* Catalytic oxidation
CO_2. *See* Carbon dioxide
Coagulants, 33, 50, 63, 67–68,
    74–75, 81, 82 f, 114
    defined, 175.
    See also Flocculants
Coagulation, 51, 63
Coal. *See* Anthracite
Cohabit, defined, 177
Colloid, defined, 177
Colloidal, 67, 102
Color, 11, 70, 89, 135
    apparent/analyzing, 142
    comparators, 65
    intensity of, 64
    perception, 64
    true/apparent, 142
Colorimeters, 63, 64, 65
Complexing. *See* Organic complexing

Compliance issues, 2, 5, 12
Compound chemical, defined, 177
Compounds, 30, 34, 177
    organic, 32, 180
    poly aluminum, 33
    toxic, 106–107
Conceptual evaluation, 42–43, 44
Conditioning, 60, 90, 96
    defined, 176. *See also* Regeneration
Conductivity, 138, 141, 142
Construction, 13, 20, 77
Contact time (CT), 47
Continuous dissolving systems, 58
Continuous regeneration (CR), 90,
    91, 94, 98, 100, 152
Conversions, general, 199–202 t
Corrosion, 4, 76, 133, 139–141,
    140 f, 147
    alkalinity and, 138
    buffer capacity and, 138
    by-products of, 133, 135
    copper, 138
    inhibitors, 137, 172
    iron, 133, 135, 139
    lead, 138
    microbial-induced, 139
    pH value and, 29, 138
    rates, 137, 138
    water quality and, 137
Corrosion concentration cells,
    development of, 138
Cost-benefit analysis, 74
    defined, 177
CR. *See* Continuous regeneration
Crenoforms, 6
Crenothrix, 6, 105, 109, 150 t,
    155–156
Cryptosporidium, 114
CSE. *See* Center for Sustainable
    Engineering
Cured-in-place pipe (CIPP), 146

www.ingramcontent.com/pod-product-compliance
Lightning Source LLC
Chambersburg PA
CBHW060405220326

41598CB00023B/3026

*9 7 8 1 5 8 3 2 1 9 8 5 0 *